Bempston

May 20

LOOKING FOR
THE GOSHAWK

LOOKING FOR
THE GOSHAWK

Conor Mark Jameson

BLOOMSBURY

LONDON · NEW DELHI · NEW YORK · SYDNEY

This book is for the dedicated souls who give their time and often show great courage and personal sacrifice in protecting wildlife. You know who you are, and I salute you.

The opinions and feelings expressed in this book are my own. This is a description of a quest for knowledge and understanding, and I won't claim that all of the answers are right. Nature does not give up her secrets easily. It has been necessary in one or two instances to change the names of people and places.

Published 2013 by Bloomsbury Publishing Plc,
50 Bedford Square, London WC1B 3DP

Copyright © 2013 text by Conor Mark Jameson

Copyright © 2013 inside illustrations by Toni Llobet

Copyright © 2013 cover artwork by Darren Woodhead

The right of Conor Mark Jameson to be identified as the author of this work has been asserted by him in accordance with the Copyright, Designs and Patents Act 1988.

ISBN (print) 978-1-4081-6487-7
ISBN (e-PDF) 978-1-4081-8701-2
ISBN (ePub) 978-1-4081-8702-9

A CIP catalogue record for this book is available from the British Library

This book is produced using paper that is made from wood grown in managed sustainable forests. It is natural, renewable and recyclable. The logging and manufacturing processes conform to the environmental regulations of the country of origin.

Page design and typesetting by Nimbus Design

Printed and bound by CPI Group (UK) Ltd, Croydon, CR0 4YY

10 9 8 7 6 5 4 3 2 1

MIX
Paper from
responsible sources
FSC
www.fsc.org   FSC® C020471

Contents

To write something which was of enduring beauty, this was the ambition of every writer ... the artist yearned to discover permanence, some life of happy permanence which he by fixing could create to the satisfaction of after-people who also looked ... Wheelwrights, smiths, farmers, carpenters, and mothers of large families knew this.

T. H. White
The Goshawk
Part 2, Wednesday
1937

Prologue

"*Goshawk*." Dave utters the word, the name, quietly. Partly to me, I sense, and partly to himself. He doesn't have to speak loudly, here. The forest is quiet. Cathedral quiet. Beyond its roof the silence – but not the peace – is disturbed only by the sporadic inhalations of breeze into forest canopy, a hundred feet or more above us. The tinkle of a burn – or stream – deadened, softened by mossy banks and soggy fallen branches, adds further mood music, easy to miss unless you stay still, as we are doing now. And listen. I think subconsciously we are listening for the Goshawk. A faint hope. The Goshawk gives very little away. Why should it? It sees like a demi-god. And hears. It moves like an apparition, guided by a rocket. It may not even be here. Except in our heads. In our heads it is vivid.

The forest floor is open, lurid with the tones and textures of fresh moss, in clumps and swathes, rugs and stair runners on both steep and gentle slopes, enveloping and enfolding rocks and wind-snapped trunks, stumps and branches in turn. This is outdoors as room. Padded. Comfortable and comforting. Mild and wild. 'Semi-natural'. *Sauvage*, in a second-hand way. I feel relaxed, at home, soothed by sylvan ambience, by the softened edges, the scent of moss, toadstools, soft earth. At the same time I am hyper-alert, sensitised. I'm in a place I've never been, with a man I've only just met, yet the sense of belonging is assured.

The roof of the forest is propped on clear, scaly, perpendicular tree trunks. These are straight, and true, and solid as the pillars of a living temple: Norway Spruce, Japanese and continental European Larch and Sitka Spruce trees; widely spaced, long-since thinned of near neighbours to promote growth over the course of their 80 years lived so far. They have been here since planting between the wars, the first phase of the modern era moves to make Britain more self-sufficient in timber, and paper pulp, less vulnerable to the vagaries of international diplomatic relations. Farmed trees. We were putting some forests back, in most cases a new kind of forest cover. We were using trees from Scandinavia, Japan (the larch) and North America (the Sitka), in the main. Of the fast-growers these are better suited to our wetter, thinner, more acid soils. They grew quicker and straighter though much less attractively than even the native so-called Scots Pine, the tree that edged north behind the retreating ice of the last glaciation to encase large areas of the northern, less hospitable parts of our isles.

It is tempting to scorn these plantations, thinking them dark and lifeless. But in maturity, like this, thinned out and given some encouragement, these woods and forests assume a grandeur and charm of their own. They are trees, after all. They will at last begin to look and feel more settled, more at home, to recapture nutrients, build soil, sustain life. The life in and around them is starting to find its feet, although the signs of human forest industry are never far away. This is not the picture postcard romanticism of the Caledonian wood, with its heather and blaeberry under-storey, its Capercaillies where these survive and its Red Deer stags. But it is proper forest in its own way. Somewhere now worth exploring, somewhere with secrets to divulge. Artifice, perhaps, with an underlying commercial purpose, for sure, but there is beauty here too. My views on this have changed. Looking for the Goshawk has helped.

"Goshawk?"

Dave and I are crouched like detectives beside a scattering of feathers at the base of one of these thick, iron-solid and branchless lower trunks. The feathers are fresh, still buoyant, alive, almost, on a bulging knoll of sphagnum moss, itself feathery, but also moist, soft and spongy to the touch. A comfortable death-bed.

Dave's terrier Hamish is sniffing the feathers closely. Some, from the victim's tail, show traces of pink-red where the tips of the shaft have been pulled, intact, from a rump. A downier breast feather is sticking lightly to Hamish's muzzle. It shivers on his breath. He makes no attempt to remove it. Perhaps he can't feel it. He looks suitably sombre, tactful. Paying his respects. He is probably taking his cue from us. Well, David.

The larger feathers – wing primaries and secondaries – are mainly brown, barred fawn and off-white. The breast feathers are more numerous, spotted and striped brown on a paler background. Some move in the vaguest air current, drifting. They definitely haven't been here long, maybe even minutes, if not hours.

They are hawk feathers. Sparrowhawk. Female. Adult. A large enough bird in itself. An accomplished predator, usually of birds. And here it is, itself depredated. No trace remains of its body.

When Dave said 'Goshawk' he wasn't misinterpreting the traces of the find we are now kneeling beside. He was naming the nemesis: the bird that almost certainly killed it, plucked it here, and took its body away to consume in secret, perhaps even at a nest. We hope a nest. But this is the only sign we have found. The only hint.

"We can rule out Tawny Owl?" I venture, after a while, taking some photos of what at this moment feels like a crime scene, of sorts. Perhaps something will come to light later, in a moment of calm reflection on these pictures.

"I think so," Dave replies. He is turning over this possibility, this suspect, in his mind.

I am prompted to ask this because just a few days ago I was with another Goshawk tracker, in another plantation forest. He told me he'd discovered evidence of Tawney Owls killing and eating Sparrowhawks – even adult female Sparrowhawks (Ian Newton first noted this in the 1970s). He has found hawk remains in some of the Tawny Owl nest boxes that he also studies in that forest. The hawks can be taken by Tawny Owls from a roost, or even a nest, in the night. The Owl – the nocturnal predator – has all the advantages in the dark. It is in its element, after all: primed, energised, able to see, of course, and silent. A daylight active raptor like a Sparrowhawk, even a large female, may know little of what has happened as the needle-sharp talons of the owl close around her catatonic head. The hunter hunted. Death will follow quickly.

That the owl – little bigger though quite a bit heavier than a Sparrowhawk – can lift such a quarry back to its nest is perhaps the more remarkable aspect. Of the larger birds, Moorhen and Snipe remains have also been found in Tawny Owl nests. Unmistakeable body parts among the prey debris trampled by young owl feet into the litter that builds up at the bottom of the nest hole. When their main prey of Field Voles is scarce, Tawny Owls have to broaden their predatory horizons, to find other ways of provisioning their hungry owlet broods.

This year is a vole year. The populations of these small and prolific grassland rodents boom and bust. We are currently at the top of a boom cycle. There is evidence of voles everywhere, and the owl broods we ringed were five strong – a lot for Tawnies. They are even hunting by day, to keep up with demand and make the most of the vole glut.

Goshawks, on the other hand, deprecate Tawnies. Tawnies cannot as a rule hunt adult Goshawks, even under the cloak of darkness. Goshawks, as much as five times the weight of Sparrowhawks (comparing like with like – male with male, female with female), are beyond the range of Tawnies, even drugged with sleep and under cover of night-time. By day, Goshawks are intolerant of other large predatory birds near their nesting territory. And that includes Tawny Owls. Disputes will occur, and there can only be one outcome. Either the Tawnies are displaced and move out, or they are killed. Dave has known Goshawks to strew the plumage of Tawny Owls in prominent places, like posting an advertisement of occupancy and intolerance.

"It reminds me of gangster films," Dave had told me earlier. "The Goshawks are advertising that they are here, to other birds. A warning. Sometimes you find the owl carcase as well, not even eaten."

Such is the life – and death – cycle of the forest. It's a bird-eat-bird, eat mammal, eat bird, kind of place. And the Goshawk has a rightful place right at the heart of it. Despite its size and power, it is not invulnerable. Goshawks, especially young Goshawks, are at risk from Pine Martens, which can readily reach Goshawk nests. Martens would strike most likely when the female Goshawk has her back turned, or has left her young unattended. And of course especially at night. Wildcats too, theoretically, although the chances of two such scarce animals crossing paths must be extremely slim.

"This looks like it happened this morning," Dave concludes, after careful mulling over of the evidence.

We can be fairly sure it's not the work of a mammalian predator. The feathers are plucked, not bitten and broken, or snapped, as happens in the teeth of a Fox or marten, even in the unusual event that a Fox or marten could get hold of a healthy adult Sparrowhawk. All things considered we can also be confident that a Tawny Owl hasn't carried out the attack on the Sparrowhawk of which we are now contemplating the aftermath, the lingering evidence. We leave the scene to explore further. "Rakin' in the wids," Dave calls it, though we tread lightly, and search gently.

Twenty metres further on I find another, smaller, scattering of feathers. Same bird. It has been taken to the cover of a low spruce, a regenerating sapling. A Tawny Owl would have little need to take its prey under cover like this, by night.

But we cannot be 100 per cent certain. Not much in wildlife forensics in the field like this is certain, it seems to me. Even with the kind of field skills I've seen in action today and with the other experts I've met and spoken to in recent days. But all the signs point to Goshawk. Dave is pleased, although he keeps his emotions in check, and to himself. I have noticed that scientists, data gatherers, people in these roles, prefer to keep their feelings under wraps, disguised.

He is entitled to share my excitement. It is Goshawk, after all, that we've been looking for. But why here, of all the patches of this extensive forest, that covers hundreds of hectares of the hills and valley sides of this part of central Scotland? Dave brought me here in his Land Rover, through locked barriers and rutted tracks, because some weeks earlier, in March, a Goshawk was seen displaying over this part of the forest. Early spring is the one time of year that Goshawks, if present, can usually – but not always – be relied on to make their presence known – provided you know what to look for. The displays are of course intended by Goshawks to impress each other – mates and rivals. Their urge to do this outweighs the risk of giving themselves away to prying human eyes, so this is when it is best to try to detect the presence of Goshawks. But even armed with this knowledge, and for all the extravagance of these 'butterfly wing' display flights, the investigator can all too easily miss them.

Goshawks prefer to display early in the morning, and for just a brief period on sunny days. Goodness knows these can be few and far between in Britain, and certainly in Scotland, at this time of year. And Goshawks tend to like remote areas, like this one, here in the Trossachs National Park. We are half an hour north of Glasgow, and west of Stirling, the gateway to the Highlands. Even roller-coastering in the sky, Goshawks are easily overlooked. Perhaps underlooked would be the word, or misidentified. Factor in too that almost no one is looking, or really knows what to look for. Why should they? The Goshawk has been a long time gone.

It is early May. If there are Goshawks here, and breeding, by now the female should be on her clutch of eggs. With my visit and offer of help, today is the first chance that Dave has had to follow up on the report, to check out the forest for further signs of Goshawk presence, maybe even find a nest itself. We have drawn a blank – no pellets, no prey remains, no shit – until we found these Sparrowhawk feathers.

Dave has been a gamekeeper in an earlier life. He has been a man who shoots for recreation, and as a professional deer stalker for the Forestry Commission. He tells me that he now gets more fulfilment from catching animals to tag and track them, to understood how these woodlands work, than from shooting. Now, he monitors birds of prey in this part of the world. Birds of prey of all kinds, along with other birds, and other animals. Water Voles, for example, have been reintroduced, and we aren't a million miles from where Beavers have been put back, after decades of enforced absence from the UK, although they aren't on Dave's beat. Much of his working time is taken up with looking after Ospreys, the fish-eating raptors that are now spreading steadily to new breeding sites, and reaching south to England, Wales and maybe soon Ireland as well. Ospreys, Peregrines, kites and harriers have occupied a lot of conservation time, effort and resource. Goshawks haven't.

Dave *has* known Goshawks to nest in the National Park in recent years, known the excitement of finding them here, thinking that the birds have returned, to stay. And he has also known the disappointment and bewilderment that comes when the birds have subsequently disappeared; adult birds and the young, fledged offspring that they have produced. All gone. Birds so powerful, with all this vacant land to reclaim, mysteriously vanished.

"They seem secure in the forest, when breeding," Dave tells me. "But in winter they move around, follow the Woodpigeons as they move to the farmland to forage. And the thrushes, as the flocks arrive from Scandinavia. And then they don't come back. They vanish."

It's a tale I will hear on several occasions in the course of this search.

Author's note

This book is about looking for the Goshawk, mostly unassisted, but sometimes with the help of Goshawk protectors, in person, in the field, and through the words that they have written. I am especially indebted to Mick Marquiss, to whose expertise, dedication and generosity I am in great debt, as I hope will become clear.

Like Mick, and Dave, and Malcolm, and other dedicated souls, working to defend birds of prey, I have been bewitched by this mysterious, elusive bird, this perceived outlaw, this historical enigma, once loved and valued, kept and trained, lost and found again, in an earlier age, then shot and trapped to extinction in Britain more than a century ago, and in conflict with thieves, trappers and gun-wielding law-breakers since its faltering resurrection.

The Goshawk regained its foothold in Britain from around the time I was born, in the late 1960s. It came back not by natural regeneration from immigrant continental birds, it seems, but mostly by escaping, or being allowed to escape, by people who kept them. Hawkers have always been abandoned by their hawks. We know, for example, that *Sword in the Stone* author, T. H. White lost his, 30 years previously. And yet, decades after re-colonisation, the species remains rare and unseen, mostly unknown.

I want to understand why the noble hawk is apparently struggling to reclaim its former place in the British Isles, a geographic zone that includes Ireland. And what exactly was its former place? Other raptor species like the Buzzard, Osprey, Peregrine Falcon and Red Kite have enjoyed a renaissance, are back in our lives, have been championed along the way by numerous people and organisations. But not the Goshawk, perhaps the most formidable predatory bird of all of these, the most raptorial of raptors, and certainly the most ambitious in its range of prey.

Perhaps I should declare an interest here, in case it isn't already obvious. I am fascinated by wildlife, and in particular how it interacts, how it thinks, and how it behaves in relation to me, to us. If I have a particular fascination with birds, it is because this relationship is easiest to observe in birds. I am conscious that it may seem strange to be especially besotted with a species that kills and eats other birds, besides a long list of mammals. And while I may sometimes flinch at the deadly ruthlessness of the Goshawk, I want to explore how its predatory presence is a vital but largely missing strand in the matrix of life in which we function. I have a strong instinct that the Goshawk has a lot to tell us about us.

In researching and writing this book I have also hoped to better understand the roots of my own fascination with this bird, and these questions. I have some half-formed sense that a mere bird can encapsulate so many things: our values, hopes, dreams, politics, history, how we see, what we don't see, what we believe, religion, and understanding of the world from which we have come – the Goshawk more so than most.

As Ken Kaufman wrote, in *Kingbird Highway*, 'The most significant thing we find may not be the thing we were seeking'. I mention this quote because although I've been looking with half an eye for the Goshawk all my life, it has only been in recent years that I've stepped up the enquiry, that the 'case is re-opened', so to speak. This happened after an incident in, of all places, a junk shop.

CMJ
Bedfordshire
Winter 2012/13

Chapter 1

JUST LOOKING

Sunday

Kielder Forest. The moorland breeze of late summer is temperate, the sky overcast, but high and dry. I am almost at the border. Almost home. I have reached a familiar, welcoming avenue, an elevated gateway to Caledonia, as it was for the Romans, and any traveller before them and since. Dere Street runs close by, the Roman road north/south. These uplands form the barrier between England and Scotland, crested by Hadrian's Wall, a little to the south of here. The battle site at Otterburn (1388) is signposted from the road; the Roman camp at Rochester too, one of two bases locally, besides 14 of their forts. It's been a busy thoroughfare through history, though tranquil today: just me and the breeze and a pipit, at a lay-by.

In the 12th century it was one of the great royal hunting forests, maintained and protected for the pleasure of the King. It is recorded that the rights to hunt "with their men and dogs with the horn, bow and arrows, without hindrance from anyone, and at all seasons of the year," were granted by the de Umfraville family in return for a payment of one Sparrowhawk per year. I wonder what they could have got for a Goshawk. I wonder if they actually *meant* Goshawk, and were confusing the two species.

This remote pass has been used over the centuries by marching armies, cattle thieves, smugglers of whisky. It has been a wild place, for wild people, and other creatures. A priest who later became Pope used this road in 1413. He noted that the local men were 'small, bold and easily roused, and the women fair, comely and pleasing, but not distinguished by their chastity'. It's fair to say it feels much less wild today. In fact it is resolutely peaceful, after five hours on the busy A1, which I left at Newcastle, with its queues of traffic snaking back to the Angel of the North, the proud, tall, iron man with wings.

After Newcastle the road home undulates and bends over and through rolling farmland, beech hedgerows, stands of mixed trees, and then steadily upwards to the border crossing point at Carter Bar. Caledonia beckons. Apart from the wind, the main sound is of bleating. This is a land of sheep grazing, with a small herd of wild goats living like outlaws on the remotest tops, and geometric, commercial forest blocks. The trees are mainly Norway's, and North America's, and the lines they form are those of between-the-wars tree planters in a bit of a hurry, and with a fondness for order.

The visitor reaching this point could be forgiven for thinking they've already reached Scotland, as the land around has long since become the uplands: scruffy, exposed and even desolate for stretches. Before reaching the border the road cuts through the eastern edge of another state-owned and managed conifer forest called Redesdale, linked to Kielder. The traveller may not realise it but they are now entering Goshawk country. Not all the wildness is gone. I like that. I won't name every place and person accurately in this tale, but I can name this one because it's no secret hideout, for a species that in all other ways is enveloped in myth and skulduggery.

This region's reputation as bandit country, populated with 'savages and robbers', endured for many years. In winter, snow often cloaks the peaks. Sometimes it barricades the road, and the border. Each time I have drawn nearer to the forest over the years of making this journey I have been alert to the possibility of seeing a Goshawk. I don't think I've ever got here without the thought of one forming in my mind. Like today, I usually stop for a bit of a recce, take the air, call to let Mum know I'm half an hour away. The tunnel of forest, which the rising road arrows through, is edged on both sides with towering larch and mature spruce, extending elegant, weeping sweeps of needles roadward in greeting, beckoning. In poorer light this place can also look dark and a little foreboding.

I am enchanted by the idea that there are Goshawks in there, somewhere. I don't need to see one to feel good about this. I like the Goshawk for its own sake. I don't want to own it, or list it, or capture it or even photograph it. But I do want to explore this meaning it has for me, and hopefully share it.

By this time I've been at the wheel for about six hours. I also sometimes stop the car here when on the way back south, even though the journey has only just begun. I first passed this way when I migrated south from Scotland, for a job in Cambridge, 350 miles away, my life packed in a beat-up Ford Fiesta. I had been living in Edinburgh's Georgian new town, in the shadow of a Kestrel-capped cathedral, surrounded by friends.

Those were fun if uncertain years, post-University, finding our way in the world, but like a passage raptor a restless Celtic spirit nagged at me and I had to move on, find a new life, a rural existence, some space of my own. Tears stung my eyes all the way to the border, right about here. Then I stopped; the tears and the car. And I looked at England, like I'm doing right now, here a long sweep of moorland valley, and it didn't seem so different, nor so far away as all that. The forest seemed a welcoming transition, somehow.

I still use the words 'coming' or 'going' home when I talk about returning to this part of Britain, although I have lived in England for almost two decades now. I was lured south by a job as a copywriter for the Royal Society of Chemistry, in Cambridge. After a couple of years there, confirming to myself why I got out of science at university, and into the arts, I got a job at the UK headquarters of the Royal Society for the Protection of Birds (RSPB), 20 miles west of Cambridge, and 50 miles north of London.

It isn't Caledonia, but The Lodge is rural, has mature Scots pines in prominent, marginal places, and lies just outside the suburban sprawl of the south-east and the English capital. My home village lies a mile from the A1, the Great North Road of the Roman age that takes the traveller to Scotland, to this border, 300 miles away. In places, this vehicular aorta has swelled to the size of an eight-lane motorway, such as when it passes cities like Peterborough, Leeds and Newcastle. But where I live now it remains clotted, a tight dual carriageway, weaving through rows of terraced houses and past old coaching inns, and punctuated by absurdly congested roundabouts.

Notwithstanding its imperfections, and its improvised layout, its bottlenecks and contractions, the A1 remains one of the major transport arteries of Britain. We have learned to live with road network cholesterol, clots, thrombosis. We are a very crowded country, after all, the seventh most congested in the world, and apparently poised to find room for another 20 million in 20 years, most of us here, down south.

Another regularly furred artery, the East Coast Mainline railway, lies about half a mile to the east of my house. I am handily placed for travelling south, or north. And while Britain remains resolutely crowded and chaotic, and inefficient in moving people around, our distances aren't great, so we manage, we muddle through.

The lowland English village I live in doesn't feature on the front of any tourist maps or in any biscuit tin depictions of rustic bliss, but it remains a mainly peaceful, semi-rural location, with open space, wide skies over flattish lands, and a surprising number of increasingly uncommon bird species clinging to existence locally. I like birds because they make me feel safe. They stand for the bits, the fragments, margins, corners and airways of the planet that we haven't yet completely subjugated. It took me a while to twig that other people don't necessarily see birds this way. Not yet, anyway.

My village is also handily placed for going somewhere else. We have another transport route here – the rivers Ivel and Great Ouse. Before the railways came, coal and other goods were taken by barge from up north, to markets and industry in the always more prosperous, more handily placed, more temperate and generally more fertile south and south-east.

Long before that, the Danes used this waterway to venture inland from the North Sea. Right here, where the two rivers collide, was the site of a battle in 797 when a Saxon army met a Danish or Viking crew on the river flood meadow, in defence of Bedford. Though not oft-cited, being so long ago, it was evidently a bit of a turning point in British history, this set-to on the floodplain. It merits a mention in Winston Churchill's *History of the English-speaking Peoples*. Some of the Vikings' top dogs fell in that wet field, that summer day. I think the river was very much wider, then, untamed, in a fluctuating landscape of water and mire.

Our village community perhaps underplayed this significance in the signs and leaflets we produced for the moated site, after we had cleared scrub and built a causeway over the moat to the island. The moat itself we couldn't touch, to dredge for restoration of the channel. The archaeologists prefer to leave these things undisturbed, other than by Badgers and Rabbits. Better to let the secrets lie, than to unlock them piecemeal, badly. Let them preserve their potential, even if we can't see it, goes the thinking.

So I like that the road through my village brings me – and all those people who've used it through history – all the way north to Scotland, and back. My parents retired from the small town satellite near Glasgow where I went to school, to an equally small town in the Borders. It still has some of its former woollen mills by a lumpy river, cradled in hills grazed by those legions of sheep and nowadays decorated mainly with those angular commercial forestry blocks. These cling in places to otherwise bald hills, nothing craggy or dramatic, but some are high enough to be classed as mountains, though grassy and rounded to their summits. Snow sticks to them for part of the winter, and the whole provides a buffer between north and south. They help make Scotland what it is, and provide the tamed and perhaps slightly misleading, seamless entrée to the country for the traveller from the south.

Taken as a whole, Kielder is the biggest forest in all of England. Some say it is the largest 'man-made' one in Europe. Ditto its reservoir. It's not all in one piece, though, and parts are even on the Scottish side. Redesdale forms part of the network of plantations across 100,000 acres. In places it is densely packed with its mainly non-native spruce tree species. Scandinavians would feel at home here. Finnish Goshawks certainly do. They have survived here in apparently healthy numbers, since they were helped to return 40 years ago. The received wisdom is that they find sanctuary among these dense stands of dark, impenetrable evergreens, little visited, barely visitable, in fact, apart from one well used, unmetalled track running east to west across it. There is something about Kielder that evidently suits the Goshawk, although such places are often derided by conservation-minded outdoorsy types for their general and in some cases rather stark and obvious lack of wildlife interest, light and variety generally.

There are usually Woodpigeons here and there, and sometimes I hear the mournful wail of Lapwings and Curlews from the moors beyond – ghostly on the wind, like Cathy calling to Heathcliffe. But in the 20 years that I have volunteered and worked for the RSPB, multiplied by the several journeys north/south per year, I have yet to get even a sniff of a Goshawk here. Goshawks are like that. They are just plain difficult to find.

I did make one serious effort to see a Goshawk here, more than a decade ago. I wrote about it for *Birds* magazine at the time, under the heading 'Renaissance'. It was intended as a modest celebration of the bird species that at that time were slowly but surely recolonising this area, following years of absence. The reason for the disappearance of these birds – species like the Golden Eagle, Raven and Buzzard, as well as the Goshawk, which was the centrepiece of the story – here as elsewhere, had been persecution. In an earlier era of zero-tolerance gamekeepering, and guilt-free trophy hunting, they had been shot, trapped and poisoned to local extinction. The hills had been empty of their calls, their soaring shapes, their ecological roles, for many years.

I met up with Dave Dick, then Head of Investigations for the RSPB in

Scotland, and Dave had arranged for us to accompany Malcolm Henderson, who is part of a small and dedicated network of volunteers called the Raptor Study Group, to a known Goshawk breeding site. There is a reasonably good chance of seeing Goshawks displaying over territories in early spring.

Malcolm was with Lothian and Borders Police at that time: a Wildlife Liaison Officer, to give him his official title. Every force has them now. Incredible as it may seem, wildlife crime is second in scale only to drug crime these days. There's the domestic stuff, like raptor persecution, and then there's the international racketeering. Malcolm is a Goshawk specialist, but this is a spare-time activity, carried out through the Raptor Study Group. This expedition was strictly a spare-time activity. We set off for the hills.

I remember Malcolm telling me that the entire Goshawk population in his wide Borders study area may originate from a single pair of birds released into the wild by a falconer in the early seventies. There was a vacant niche that the species was reclaiming, an important role in what is loosely called the natural order of things. There is no shortage of Woodpigeons, crows, Rabbits and Grey Squirrels here or in most parts of Britain. The Goshawk is better equipped than any other native predator to regularly take prey of this range and size.

But apart from a nest, the odd pigeon sternum and Rabbit thigh bone, there was no sign of the birds. I remember how it felt like they knew we were here; like we were part of the platoon in the film *Southern Comfort* – outside of our comfort zone, and inside someone – *something* – else's. Well as Dave and Malcolm know the terrain, something out there was a step ahead of us – knows it better. Luckily for us, Goshawks don't lay traps.

A local shepherd told us he'd seen the Golden Eagles around. He knew they were trying to come back. Maybe this year, maybe next, he reckoned. We were just a short hunting flight from the border, and the open spaces of north England beyond. The shepherd seemed quietly proud, and told us he was keeping an eye on things.

We saw all of these potential prey species that day, but no, I didn't see a Goshawk. I saw their home, but I didn't see the occupiers. But I genuinely wasn't disappointed. I got to know the species better, because the experts had shown me round its neighbourhood. In many ways the fact that the bird eluded our eyes only added to the mystique.

* * *

Apart from that one outing over a decade ago, my efforts to find the Goshawk here have been cursory, not sustained or systematic. And despite that visit, the bird had still not been fully formed in my consciousness. Although still haunted by it, I suppose you could say I still didn't fully believe in it. And Kielder is too near the end of my journey north for me to want to or to be able to spend much time searching; just as it is too near the start of my journey south, on the return leg, to do likewise. But I have always thought about it, here, like a man who cannot pass a chapel without blessing himself, although he's not regular in the pews. Perhaps *because* he's not a regular.

Goshawks manage to be as elusive as eagles, or Foxes, but without resorting to living in the mountain tops, or moving mainly by night. They are highly sentient – one saying goes that they can see through walls. They will certainly detect your approach, by sight or by sound, from a mile away, or more. They have learned over the years, by direct experience or from the reaction of their parents when they are young. that humanoid forms equal trouble; are best avoided. It's a sensible strategy. "We reek of death" after all, as author J. A. Baker put it.

But you'd still think in all those years I might have had just the tiniest glimpse of one, caught unawares, careering through the trees, firing over the road, perched on a spruce tip. But no. Nothing. Not yet.

Saturday

It is when I least expect it, am not even thinking about it, heading south, Kielder two hours behind me again, that I find the Goshawk.

I'm in a town in northern England. Let's call it Lamberton. I have found this town a convenient place to break a six- or seven-hour journey. And I have old friends here too, and I sometimes stay with them.

I've parked the car and am passing a warehouse in a side street, just outside the town centre. It might be the bust of Elvis that catches my eye, or the old jukebox, but I am enticed inside to have a browse around. I'm hoping, perhaps to find, as T. H. White put it, something of enduring beauty.

Some things get a new lease of life here. Perhaps that's why I am drawn to junk shops, charity shops even better. I like rescuing things; to find and promote the enduring beauty. I think of these shops as living museums, where you can handle and take home the exhibits.

Stuff is piled on top of stuff, as though in a very large, slightly mouldering attic. Fertile ground for turning up little treasures, curios ... things for salvaging and breathing life back into, to rub and find magic within; to polish, and thereby, perchance, to release the past.

The smell of dust and must transports me to the attic of my subconscious. I am once again eight years old, lifting the trap-door on the attic at home in west central Scotland, near Glasgow, distant in time and space, creepy and damp, and draughty, dark and cobwebby. When my sister Brigid and I came to realise that the boxes up at the end were full of surprises, and interesting old things wrapped in limp, gritty newsprint, we were more prepared to run – or crawl – the gauntlet of this gabled tunnel wherein lurked beasts, such as spiders.

That attic was a pirates' cave of such discoveries; the trappings of the past. The cardboard boxes had bags of stamps from countries I'd never yet heard of, cigarette card albums, comics like the *Victor* and *Dan Dare*, toy soldiers, tins of coins, family photos curling like fallen leaves, cast-off clothing, moth-balled and cobwebby, heavy, solid and intact.

I am once again the child in the attic, when I notice some taxidermy specimens out of reach on top of a mound of cabinets and boxes stacked in a dingy corner, halfway to the roof. Some are in cases. Taking a step-ladder,

I climb carefully, now imagining I'm in a post-modern Tomb of the Kings. I peer into a glass case, finding an angle on it to get a clear view of its contents, correcting for reflection. What confronts me is the beady glare of a lean, contoured, elegantly streaked, stout-chested and imperious bird of prey.

Looming over me, it looks larger than its two-feet of length, given added stature by imposingly powerful thighs. Its wings are partly drooped, or mantled, over the prone body of a Magpie, which it dwarfs, and on the breast of which the raptor's arched talons are emphatically clamped. The bird is muscular of chest and thigh, but beautifully streamlined, long-legged, with thickly feathered tail coverts, stout shins, and breast feathers decorated with chocolate-brown arrowheads, as though dabbed with an artist's brush. There is a hint of saffron about the feathers, and the flecking of chocolate-brown is in places also star-shaped, faintly dazzling.

She is glaring at me, or through me, and to the world, through surprisingly authentic and piercing yellow-irised glass eyes. Defiance radiates from her, even in this caricatured state. It is most unsettling. Enervating, dare I say. I have eyes only for this piece of junk.

It stirs another memory, of first discovering birds like these – this species in particular – in books, as a child. Somehow this stuffed skin has retained that ferocity and intensity that struck me at a young age when I first encountered these birds of prey on the page. It made me fall for them then, in a way that was bound to last. In a strange and perhaps perverse way I am reconnecting with the Goshawk after decades of estrangement. In a junk shop. Even in this semi-parodic state, and pose, the bird I'm now looking at is beautiful, with a heavy hint of menace. It is tempting to see anger here, laced with fear, and the noble bearing reflected in the Goshawk's modern day scientific name: *Accipiter gentilis*.

This is a piece of natural engineering perfection, a bird beyond birds, a bird with an almost human expression, a furrowed brow that says something back to me about me, about us. It is a look that has haunted me ever since, and held me in a kind of spell. A *chimere*, as they say in France: both a fanciful notion, a semi-fantasy, and a mythical beast with the properties of several beasts rolled into one. Bird with mammalian qualities; with presence, but seldom seen. They might have had the Goshawk in mind when they came up with the concept of the *chimere*. They have two names for Goshawk there: the *cuisinier* (kitchen provider) and *rameur* (rower – from the way it flies when hunting).

Her partly-open, loosely draped wings are tawny brown with lighter edging, emphasising the shape and texture of this cloak. Her lethally hooked beak is slightly agape. She is undeniably magnificent, even in this inanimate state; this frozen, partly fictionalised moment in time, of predator on prey. I am fixed by her dead stare: fixated, quite possibly.

"What a beast," I'm mumbling. She is a juvenile – this plumage is very distinctive in Goshawks in their first year. Juvenile, but already sexually mature, of breeding age. And undoubtedly female – she is very much bigger than the male. Very much bigger than I'd properly grasped before. Size isn't everything, in nature, but it does count for a lot where the awe factor is concerned. We aren't most of us enthralled by elephants and whales for nothing. We think of Magpies as large, predatory, menacing birds. In her grasp, her magpie quarry looks unreasonably mismatched. I think awe – or fear – makes us see things as bigger than they are. Think rats. Or spiders. And for now, think Goshawk.

You might say I have been slightly derailed by encountering this Goshawk. Of finding this body. It has challenged my senses and my understanding of things. The hawk's gaze has pierced my soul, seen right into me, yet somewhere beyond me. Goshawks have this effect: a kind of serious intent, even this one, stuffed and mounted. It's a kind of haunting. I haven't been able to shake the impression of that bird. It has left me needing to understand how it came to be trapped in the glass case, and when, and why. And if not this one, then those of its kind. And where else is it now, this species, in its living form, if anywhere? Can we really still be oppressing something so apparently powerful, so beautiful, so inspiring, in modern-day Britain?

Who killed the Goshawk?

I've been interested before, even looked before, but from this moment I need to know and understand this bird. I am not starting from scratch in this. Far from it. I have known for a long time that the Goshawk became extinct in Britain, partly because many of them ended up in the same state as the one I've just found – trapped, shot or poisoned – and, if recovered in time and in one piece, stuffed for trophies or sold as souvenirs. It's what we did in this country, in Victorian times and earlier. Even since.

It is easy to be scornful now of these bygone practices, belonging to another age, another mindset, and to think taxidermy specimens ugly and unnecessary. I try to take a more forgiving view; to imagine an age when perhaps people assumed the supply of wildlife to be limitless, if they had to think about it at all, their country great, their empire greater still. They had some justification in thinking that the best way to get a decent look at something, to try to know more about it, to appreciate its beauty in a lasting way, to share it with others, might be to kill it, and bring it home.

This stuffed bird is certainly the best view I've had of a Goshawk. Of course it is. It may not be moving, but it is in three dimensions. And close. This is not to completely exonerate Victorian or even later Britain for its blood-lust, acquisitiveness and short-sightedness, but it is to acknowledge that the motives haven't all been merely destructive, vain and materialistic.

The early Victorians shot and trapped Goshawks for other reasons, as perceived opponents of game birds, for example, until they were all gone. Of five bird of prey species wiped out in the UK in this period, the Goshawk is believed to have been the first to go. At least within these islands. And all the while birds like this were presumably hated by their game-preserving persecutors, they have remained popular with others, who have kept them captive and tried to master them for falconry. It is from these escaped captive birds that new, wild, refugee populations of Goshawks became re-established in Britain's quieter state forests, such as Kielder, mainly from around the time I first encountered these birds in books, in the early 1970s.

The exhibit in front of me may have been a captive bird. I hesitate to say pet, as I think to call a captive Goshawk a pet would be slightly misleading. If it was killed just for the purpose of stuffing, in a more recent age, I would not be so forgiving.

I'm wondering if the shop has got the paperwork to legally keep, display and sell stuffed wildlife. Protected species, like the Goshawk. There is a form called an A10, issued by the Department of the Environment, which would validate the origin of this bird. Without it, a non-antique stuffed Goshawk would be illegal to keep or sell.

This one really doesn't look particularly antique. It's not the usual jaded and faded Victorian artefact you normally find in these places, like sunblanched and moth-eaten road-kill, lumpy and misshapen; like some of the other items here, today. No, this is a professional, more recent job, even allowing that the glass case must have preserved it well. It is almost tasteful. If it comes with the necessary form to show that the bird has been sourced in an acceptable way – died of natural causes, or a captive and licensed bird – and if it didn't cost several hundred pounds – what some people ask online for a legitimately stuffed Goshawk – then I might even have been tempted to take it home. Give it new life, of a kind. Show it to others. How else after all, even today, can you introduce people to the wild Goshawk?

"The stuffed animals aren't for sale, mate," says the man in the overalls down below me. The spell is broken for a second.

"Ok. No worries," I mumble in reply.

"I'm just ... looking."

* * *

I've been just looking ever since.

Chapter 2

THE MADNESS OF MR HUDSON

'When men die and go to hell, they are sent in large baskets-full to the taxi-dermists of the establishment, who are highly proficient in the art, and set them up in the most perfect life-like attitudes, with wideawake glass eyes.'

Wednesday

A solitary oil painting hangs above a sandstone fireplace in the main meeting room at RSPB headquarters. It dominates the room, which is high-ceilinged and oak-panelled, with heavy doors and tall, iron-framed windows. These face onto a formal garden and Victorian swimming pool, stone-flanked, its cloudy surface furrowed by the scaly backbones of carp. A Moorhen croaks unseen from somewhere in its midst. Jackdaws flap around and occasionally somersault overhead between chimney stacks and the sweeping limbs of Lebanon Cedars.

I'm looking again at the painting, which is of course familiar to me, and to everyone else who uses this room regularly. But I'm not sure I've properly studied it before. It depicts a grey-whiskered gentleman crouched on a grassy knoll, in a heathland setting, on the edge of an oakwood. It could almost be the Sandy Warren heath, just beyond the window pane and gardens outside. But the man in the painting, while matching the Victorian period and mood of the house and gardens perfectly, never lived here, and the RSPB itself only arrived in 1961, when it moved out of its small London offices in search of affordable room to grow, to meet the challenges about to be neatly encapsulated by the imminent publication of Rachel Carson's *Silent Spring*.

That was just over 50 years ago. This painting is almost a century old. Its subject is nattily dressed in tweed plus-fours and a cap. It's a worker's cap, but he doesn't look like he's just popped out from some dark satanic mill on his lunch break. He is clutching a small pair of field glasses, of the type that turn up in junk shops sometimes, or in Oxfam display cabinets. They appear better proportioned for use at the opera than in the field. If pressed I'd say he looks a bit like Lenin, this man. He is solemn of expression, a little grave, brow furrowed; a Victorian patriarch, no doubt – a man wearing heavily the cares of the world.

This is William Henry Hudson: W. H., to those who know him and his work. The RSPB tends not to deify individuals and few portraits are obvious here (there are none at all for any of the 'formidable women' who founded the organisation), but W. H. earns his place above the fireplace as he was around at the time the organisation came into being, in 1889. He was a staunch supporter and advocate in an era when conservation was a germinating idea in the minds of a vexed and determined minority. He was the figurehead who took tea with the ladies who did the work of setting up the Society for the Protection of Birds. No one need be in no doubt that it was the ladies who did that – hundreds of them. With the organisation up and running, Hudson was even invited to become the fledgling Society's Chairman. But by common consent his talents were better suited to other things, namely writing. He lasted just a year in that office, but whatever Hudson's shortcomings as an administrator, he was a true pioneer of conservation.

Among the many things I've salvaged from junk shops, auctions and the like is a collection of old natural history books. More of these than I can ever hope to read thoroughly now line the shelves of the bookcases also picked up from auctions or the homeless charity Emmaus, at its base near Cambridge. I enjoy dipping into some of them for snippets on subjects of interest, to inform an article I'm writing, or to answer a question prompted by an observation in nature. The old books are often elaborately, ornately written and passionately expressed, if not always altogether reliable. But what they rarely fail to convey is their writers' enthusiasm.

I have a clutch of W. H. Hudson titles. His passion and compassion radiate from every one. In scanning their indexes for references to Goshawk, I find one in *The Illustrated Shepherd's Life* – his account of the old ways of the Wiltshire countryside, in central southern England. It supplies a few gems of insight in what is otherwise pretty barren terrain for Goshawk details:

"The larger hawks and the raven, which bred in all the woods and forests of Wiltshire, have, of course, been extirpated by the gamekeepers," he wrote. "The biggest forest in the county now affords no refuge to any hawk above the size of a kestrel. Savernake is extensive enough, one would imagine, for condors to hide in, but it is not so. A few years ago a buzzard made its appearance there – just a common buzzard, and the entire surrounding population went mad with excitement about it, and every man who possessed a gun flew to the forest to join in the hunt until the wretched bird, after being blazed at for two or three days, was brought down."

It's a reflection of the prevailing attitude towards raptors, and birds generally, of the period. This is the context in which Hudson was operating, which makes his tenacity and perseverance as an early conservationist all the more admirable. He was determined to do what he could to try to stem this tide; and this a man who was born in the USA and raised in Argentina, and who only arrived to live in England in adulthood.

"I heard of another case at Fonthill Abbey," he goes on. "Nobody could say what this wandering hawk was – it was very big, blue above with a white breast barred with black – a 'tarrable', fierce-looking bird with fierce, yellow eyes. All the gamekeepers and several other men with guns were in hot pur-suit of it for several days, until someone fatally wounded it, but it could not be found where it was supposed to have fallen. A fortnight later its carcass was discovered by an old shepherd, who told me the story. It was not in a fit state to be preserved, but he described it to me, and I have no doubt that it was a goshawk."

Hudson was writing in 1910. He obviously cared a lot about this bird but it's a safe bet he had never in his life set eyes on a living Goshawk. It's a northern hemisphere species and therefore doesn't occur in Argentina, so he can't have seen one there either. By common consensus the species had been driven to extinction in the wild in southern Britain in the second half of the previous century, perhaps even earlier. It says much about his knowledge and open-mindedness that he can have speculated that a Goshawk was the bird in question. If indeed he was right, and the bird was wild in origin, it is likely to have been a passage bird, visiting from the north of Europe. Goshawks drift south and west from there in autumn and winter in varying but low numbers each year, occasionally reaching or passing through Britain and spending time based in woodlands here.

These passage birds are thought unlikely to remain in Britain, and in-variably to return to the continent in time for the following spring. Passage birds are thought therefore unlikely to have been involved in the Goshawk's post-war recolonisation of Britain. It is one of those things about which ex-perts appear to be sure, but which depends on some fairly major assumptions about Goshawks always adhering to a particular pattern of behaviour. What we know from other birds is that within a given species there are pioneers, perhaps malfunctioning from the norm in some way, that break the mould and enable species to colonise, and presumably then recolonise, new lands. More visible species, like Bee-eaters, breed here sporadically, to cite one ex-ample. But immigration was not, it seems, the source of today's Goshawks.

Another of Hudson's books, and this time not a recent edition, catches my eye as the image of the Goshawk in the junk shop continued to haunt me.

My copy of *Birds and Man* is one of those weighty old tomes, tatty-edged, in a faded green hardcover with gold lettering indented on its spine. It was published in 1915, during the First World War, having been, it says, "out of print for several years". It was mostly written about 20 years earlier. This copy has just one colour illustration, a print of a Dartford Warbler perched on gorse, glued onto one of the blank leaves at the start.

As ever, I duly look for Goshawk in the index, and am disappointed to find none there. It turns out to be an editorial oversight. Chapter 1 is headed *Birds at their best*, and from the first line I am already finding what I'm looking for. This is Mr Hudson's take on taxidermy.

"Years ago" he writes, "I spoke of the unpleasant sensations produced in me by the sight of stuffed birds ..."

He warms to the theme:

"When the eye closes in death, the bird ... becomes a mere bundle of dead feathers; crystal globes may be put into the empty sockets, and a bold life-imitating attitude given to the stuffed specimen, but the vitreous orbs shoot forth no lifelike glances; the 'passion and the life whose fountains are within' have vanished, and the best work of the taxidermist, who has given a life to his bastard art, produces in the mind only sensations of irritation and disgust."

Hudson was writing at a time and of a practice that was industrial in scale and causing the deaths of countless birds, almost entirely for the purposes of stuffing and mounting for ornament. It was an industry that he clearly felt was at odds with developing a true appreciation of nature in its wild state. His writings suggest a man isolated, and up against overwhelming odds, in an age when war of a kind was being waged on nature, untrammelled by legislation.

To love natural history was not an uncommon thing in the Victorian era, but the extent to which Hudson's concern about these trophy-hunting excesses was shared by others is hard to gauge. To most people, birds and other wildlife trophies were supplied from a source of limitless abundance.

To others, God's creatures were simply there to exploit and to harvest as we saw fit, or to eliminate as vermin if their interests were deemed to clash with our own, notably the rearing of game species.

So the RSPB was born out of this period, and specifically because of the anger felt by that brave alliance of women towards the trade in bird plumage for the fashion industry. Hudson was writing shortly after the Society for the Protection of Birds (it became Royal through its Charter in 1904) was first formed, in 1889. Attitudes were beginning to change. The SPB's membership grew rapidly to 5,000 – virtually all women.

Another chapter heading in *Birds and Man* catches my eye: *Something pretty in a glass case*. And here, with a rising inner glow and even a bunched fist, I find a Goshawk. Somehow it didn't make the index.

"Who," asks the author, "walking by a riverside, does not experience a thrill of delight at the sudden appearance in the field of a vision of that living jewel, the shining blue kingfisher! This is one of the favourites of all who desire to have something pretty in a glass case in the cottage parlour, in room of the long-vanished pyramid of wax flowers and fruit. It is, however, not only the common people, the cottager and the village publican who desire to possess such ornaments. You see them also in baronial halls. Many a time on visiting a great house the first thing the owner has drawn my attention to has been his stuffed birds in a glass case: but in the great houses the peregrine and hobby, and goshawk, and buzzard and harrier are more prized than the kingfisher and other pretty little birds."

It is not hard to picture Hudson being shown these trophies by their proud owners, and responding through gritted teeth. Perhaps he chose his words, his moment, carefully each time. Maybe the subject was broached again over port and cigars later in the evening. I wonder how often the acclaimed author was asked back to these baronial halls, if his emotions on the subject could be kept in check no better in the moment than as he poured them out onto the printed page.

Hudson is a heroic figure to me. He is before his time, a man of the world, an outsider who came back from the Americas to explore his ancestral homeland, to explore its byways and its people. It is this sense of him as a visitor that is the essence of the man. He could see our world objectively, with detached clarity. But I'm also trying to view the world through Victorian Britain's eyes. How else were people – many of them no doubt nature lovers – to be impressed by a Goshawk, in an age before widespread use of any kind of wildlife photography, let alone moving images, and very few affordable books for the masses, other than to have a dead one atop the piano, or a live one tethered to a bow perch in the garden?

"The Philistine we know is everywhere and is of all classes," our hero declares.

I can feel that penetrating gaze: *J'accuse!*

Taxidermy is a craft with ancient origins, in the curing and tanning of animal hides and skins, and then the later recreation, by stuffing, of inanimate versions of the creatures themselves. But it was not until the turn of the 20th century that its practitioners became better skilled. For there is evidently more than one way to mount a specimen. I might cite this in my defence at having found the Goshawk in Lamberton so visually alluring (with certain caveats), given the twinge of guilt that Mr Hudson has now made me feel. I should add though that he even goes so far as to say that "the more cleverly the stuffer has done his work, the more detestable is the result". But I am just much more inclined, with the benefit of distance and hindsight, to find some element of forgiveness and extension of benefit of doubt towards our ancestors. Who knows what we might, had we lived then, have done ourselves, in the circumstances. But if I'd known Mr Hudson I am sure I would have shared his revulsion; the fire in his belly and mind.

He goes on, in this chapter about pretty things in cases, to describe an extraordinary night he spent at a house somewhere he calls a "lonely, melancholy coast". On this fervent evening the wind was swirling outside, battering the house from all sides. He was clearly unsettled by "its ravings" and "perpetual, insane howling and screaming". He describes sitting by the fire in a room with two mounted specimens regarding him from behind glass in an alcove to the side of the fireplace. One is a Green Woodpecker, the other a Red Squirrel. This was before the Grey Squirrel, first introduced from North America to Scotland in 1842, had conquered the lowland and south of Britain, out-competing and sweeping aside the smaller reds before it.

In an extraordinary account, Hudson explains how he began to imagine these two creatures having a conversation, turning into words the sounds of the gale outside. The pair quarrel at first, then describe how they came to be in this state, shot in a local wood. Their attention then turns to the author.

"Do you know," says the squirrel, "I think he is going mad. He should, in his raving madness, snatch down our cases from the niche and crush them into the grate with his heel."

Mr Hudson admits that he might be going insane. "What wonder that, when hours later I fell asleep, I had the most distressing and maddest dreams imaginable!

"One dream was that when men die and go to hell, they are sent in large baskets-full to the taxidermists of the establishment, who are highly proficient in the art, and set them up in the most perfect life-like attitudes, with wideawake glass eyes, blue or dark, in their sockets, their hair varnished to preserve its natural colour and glossy appearance. They are placed separately in glass cases to keep them from the dust, and the cases are set up in pairs in niches in the walls of the palace of hell. The lord of the place takes great pride in these objects; one of his favourite amusements is to sit in his easy-chair in front of a niche to listen by the hour to the endless discussions going on between the two specimens, in which each expresses his virulent but impotent hatred of the other, damning his glass eyes at the same time relating his own happy life and adventures in the upper sunlit world, how important a person he was in his own parish of borough, and what a gorgeous time he was having when he was unfortunately nabbed by one of the collectors or gamekeepers in his lordship's service."

We can't, of course, find him preserved in his own glazed, dust-proof case by the RSPB fireplace, but we do see this almost life-size portrait. And while he doesn't speak from within this frame of his adventures in the upper, sunlit world, his resolute, faintly frowning gaze – like that of the glazed Goshawk – speaks volumes yet, for anyone prepared to listen.

Sometimes, if I am giving a presentation in this meeting room, I make reference to the man above the fireplace. If I am talking about raising funds for conservation, I speak of Mr Hudson's £6,000 legacy to the RSPB – a life-changing donation for the organisation back then, and which would be a six-figure sum today – the difference between a project succeeding or failing. He asked that the money be spent on education, and within two years, but the redoubtable Mrs Lemon, who ran the organisation at that time, and knew a thing or two about prudent administration, persuaded him to relax the conditions on his donation. In doing so he would be helping far more. He ought to have known this from his time in office. But then he was a writer, and a romantic, not a bean counter. His image hangs here (while the ladies' do not) because of the terms of a legacy.

Mr Hudson remained on RSPB Council, an *éminence grise*, until his death in 1922. He survived just long enough to witness Parliament's passing of the Importation of Plumage Act, after more than 30 years of campaigning by the now Royal SPB. Conservation campaigning was ever a long game. Standing before the painting of Hudson I read his furrowed brow now as care-worn. The quoted paragraphs above suggest he was just that, simmering, and at times demoralised. I think that in his shoes, in his day, I would have been too. But with the benefit of hindsight, with 100 years distance, I can afford to be more forgiving of his contemporaries. On balance, I'm guessing that if I hadn't owned field glasses, and books, or even been particularly literate, and hadn't understood wildlife as a finite resource, I might have wanted some pretty things in glass cases too. And even the odd beautiful, fearsome 'tarrable' thing. Luckily for me we live in a more enlightened age.

The heath that Hudson is sitting on in that painting is in the New Forest, Hampshire. It was one of his favourite haunts. Another four decades would pass after his death before the RSPB moved from its London base to the house here at Sandy. I am sure he would have approved of the heathland restoration work that is going on here today, to entice back birds with which he would have been very familiar – Dartford Warblers, Nightjars, Woodlarks, Tree Pipits and Hobbies. We catch tantalising glimpses of these, from time to time. And sometimes – just sometimes – there are even reports of a Goshawk.

I should here provide a few basic facts about the Northern Goshawk, as it is more properly known to distinguish it from other, related species. For the purposes of my search I will call it simply the Goshawk. It is the biggest goshawk species and the only one that occurs in the northern hemisphere, and it is found right across the top half of the globe. It varies in size and colouration from place to place, and some observers split it into about ten subspecies, the exact delineations of which are inexact and still subject to some scrutiny. But broadly speaking there are smaller, darker ones on Mediterranean islands, and much bigger, paler (varying to almost completely snow-coloured) in Siberia, and all the variants in between. Eye colour ranges from pale yellow to blood-red, usually getting darker with age, but not always. Red eyes seem more prevalent in North American birds. The female is always substantially bigger than the male. She can be two feet or more in length, have a wingspan of more than four feet, and weigh more than 1.5 kilograms. This is roughly Buzzard-sized, although she is very much heavier than a Buzzard. The male reaches around 20 inches in length. In their first year of life Goshawks have a very different plumage to that which will follow their first moult. They will capture and eat the flesh of almost anything up to and even well beyond their own size.

The Goshawk has been dubbed 'the phantom of the forest'. Its evasiveness is legendary, to the small number of people who have got to know it, or tried to get to know it. But even allowing for this fabled elusiveness, with 500 staff and volunteers at The Lodge, if a Goshawk is around, you'd expect us to know about it. From time to time someone will report having seen one. In my close on 20 years working at The Lodge, Goshawks have been reported from the neighbourhood at the rate of about two per year. I notice the reports, if not the birds.

It has been customary that, when someone reports seeing a Goshawk, no one really believes them. It is something of a test of credibility to even risk reporting it.

Most records of Goshawk proffered to the county bird recorders are considered closely then dismissed. "We reject more Goshawk records that any other species," my colleague Richard Bashford has told me. Richard is secretary of the Bedfordshire Rarities Committee. "They are regularly claimed but most often rejected due to inadequate description. We need more detail than just 'bigger than Sparrowhawk'," he added.

Recorders like Richard are the recognised experts who compile annual lists of what has been seen (or maybe just heard) in the county. These reports form important historical records of changes in bird numbers over time. With Goshawk records submitted, it is, the feeling goes, in the end too easy to confuse the Goshawk with the much more common and widespread Sparrowhawk. In both these species, the female is bigger than the male. Usually. Things get complicated by the fact that there isn't such a huge differential between the (larger of the sexes) female Sparrowhawk and the (smaller of the sexes) male Goshawk (although male gos are substantially heavier). And isolated birds with nothing but sky around them are difficult to get a scale on.

There is a further complication, according to the theory. There are extra-large female Sparrowhawks, and unusually small male Goshawks. If a hawk is seen soaring, it will tend to be distant. If it's up close, in flight, it won't be hanging around. That's the received wisdom, anyway, but in reality very few UK-based ornithologists will have had much close-up involvement with Goshawks, or experience in getting to know the species. Most of what we know, or think we know, is second-hand, passed down from book to book. It is often based on observation of Goshawks displaying, which is when most observers have seen them, but isn't typical of the birds' behaviour or flight patterns at other times in the year.

Providing good evidence of a Goshawk is clearly hard to do. Most people who submit records simply don't provide enough of this back-up description. They could even copy it out of a book, but usually they don't bother, and therefore just don't satisfy the panels that the bird they have seen can't have been a Sparrowhawk. Or even a Buzzard, which is similar in size to a female Goshawk. Record books have to be about near or complete certainty. Reputation helps, although I've never quite worked out how an ornithologist earns a reputation – are there exams? Eye tests? There are no ID tests or spot-checks in ornithology. You'd think it might all be a bit more scientific ...

The RSPB is full of experts – world authorities, in some cases. They tend to know their stuff. They have to. They are tested most days, one way or another. Sometimes in courts, or public enquiries, by aggressive cross-examiners, by some people willing them to be wrong. They tend often to be very good at identifying things – birds and other nature. One of these colleague-friends is Duncan McNiven, who works in Investigations – wildlife crime stuff. Duncan reckons he saw a Goshawk flying over the heath at The Lodge one lunch time, a few years ago now. He described its size, muscularity, barrel chest. Because he's who he is, I think we could all take it that he almost certainly did see a Goshawk. I believed him, at any rate. He's seen quite a few Goshawks – more than most people. Most of them dead, sadly, wrapped in bags, stored in the freezer, brought as evidence in court cases, sent to museums to become skins. The live ones he's seen have mostly been seen abroad.

His report of the live one at The Lodge had made me wonder retrospectively, optimistically, a little wistfully, if I'd seen one too, in the same place. It recalled an occasion on the same bit of the nature reserve when I was looking at a Dartford Warbler that had come to spend a few weeks with us. I can place the event exactly, because I wrote about the warbler at the time. It was October 2004. This rare bird for Britain – it was probably common in Mr Hudson's day, in his New Forest haunts – had turned up unexpectedly to live on this little remnant of heath we have preserved here – a heather and bracken-covered slope, in a clearing among the Scots Pines and birches. While watching it (I was chatting to James Cadbury, long-serving RSPB ecologist), I noticed a raptor passing quite low along the crest of the heath, just above the pines there, and I blurted, "Hey look, a Buzzard!"

Back then it was still a little bit unusual to see a Buzzard hereabouts. It was certainly worth remarking upon. Buzzards made a comeback locally in the 1990s after decades of absence. They had been pushed back, since Mr Hudson's day, to the north and west of the British Isles. In recent years they had started to return steadily south and east. It seems people had stopped killing them so much. The idea goes that there were fewer people actually employed to kill them – gamekeepers, in the main. And the birds were better protected, and anyone else with a gun and a penchant for pot-shots at interesting flying things was generally becoming more enlightened, or more aware of the law, the penalties for transgression, and the greater likelihood of being caught.

A Buzzard in Bedfordshire in 2004 was certainly worth pointing out to the people around you. And someone else said, 'I think it's a Sparrowhawk.' And I, realising that it was, on second looks and thoughts, hawk-shaped, felt a bit sheepish. But I realise now that this bird that we saw was managing to look like both a Buzzard (in size) and a hawk (in shape). And it wasn't that far away, beating and gliding along the top of this bit of heath, on the edge of the pines. In roughly the same place Duncan – the more reliable authority – claims he saw his. At the time it just didn't cross my mind that it might have been the bird that is a bit like the combination of Buzzard and Sparrowhawk. The Goshawk. Not even nearly. But I never forgot the moment.

Duncan's report made me re-consider. But I would just have to add this to the short list of occasions in my life when I might have seen a Goshawk, but probably not. At least not definitely enough to satisfy the demands of the County expert records panel in my own mind.

The RSPB has a Wildlife Enquiries Unit, dedicated to fielding calls and written enquiries from the public. They've been receiving increasing numbers of photographs. It reflects an age in which most people carry telephones, and most of these have cameras built in to them. One of the things that people are photographing is hawks, and it turns out that they are often asking if the hawk in question is a Goshawk. Perhaps it's because when you actually get to have a good look at a Sparrowhawk, and this is usually when it is on a catch, or resting on the garden wall, it looks quite unlike the bird that zooms around the place, or glides overhead, and which people are mostly overlooking. Bigger, fiercer, and, yes, a bit like the Goshawk in the guidebook. Especially if it's a female Sparrowhawk, and there's no way to get a proper scale on it.

Our enquiries staff are pleased when one of these photos does turn out to be, in fact, a Goshawk. The picture was sent in from Caerphilly in Wales. Not the last place you'd expect to see a Goshawk, and it may well be a resident bird that has been captured in the lens. It is pictured standing by a yew hedge, on a rug of strewn pigeon feathers, and fallen leaves. It is in full adult plumage, and it is very distinctively a Goshawk. A male, I would say. Big, but not huge. It has a very bold and distinguished white eyebrow, thickening towards the back of the head. Its eye is orange, determined. Its crop is bulging with the meat it has just consumed. It has a silvery appearance beneath, white base colour flecked with grey, and its back is grey-blue. A beautiful bird, to my eye and to most people; vivid against the deep green of the yews behind it, suddenly present in its domain, like a nocturnal beast unexpectedly exposed by daylight.

"It made a welcome change to say, 'Yes, it is a Goshawk,'" Richard James of the Unit told me.

Saturday

The gathering autumn. The urge to collect, to hoard, to store, grows ever stronger with the shortening days. My instinct has always been to find and gather books, to salvage them and line them up on the shelves on the implicit hope that one day they will keep me. Keep me occupied, sane, informed, learning, at least; that their nourishment might be imparted to my mind. I had a spell of depression when I was ten years old. I don't think they really knew what to make of it, what to do with me. I remember one early morning, me sleepless with dejection, and my dad suggesting books. Read more books. He may have been on to something. 'The best thing for being sad is to learn something,' says the wise wizard Merlin, in *The Sword in the Stone*. Dad left to go to the school where he taught.

I live in hope that one day I will have the time and mental energy to do all this reading I've amassed, perhaps much as Hudson's squirrel or a Jay hides nuts and acorns in a woodland, in the hope of surviving winter.

Tuesday

New York wasn't always high on my long list of places to visit. Woody Allen once described himself being "as two with nature". Many of his comedies have dwelt on themes of the alienation and neurosis of life in New York City. Watching these, when I was a film student at Stirling University, New York seemed to me a lawless, dystopian place. It felt light years from my home among the woods and hills of rural Scotland, and about as far removed from nature as you could get. I wondered if, rather than therapy, it was a big dose of nature that Woody's characters needed.

There is a name for this phenomenon now, growing in prominence: Nature Deficit Disorder. It was a term coined by the American academic Richard Louv. He wrote a book called *Last Child in the Woods*, about the damage that might be done to children's development by their increasing disconnection from nature while growing up. No such theory existed back then, but I think I already had a sense of it. On the testimony of Allen, and Martin Scorsese, another of the *auteurs* I studied, late eighties New York would not be high on many people's list of places to visit. By 1990, the city's budget deficit had sunk to 1.5 billion dollars, and life for the city's residents was about as tough as it gets.

"Some day a real rain is gonna come, clean up these streets ..." promised Travis Bickle in Scorsese's *Taxi Driver*. Until it did, I was staying home. I forsook Hollywood for conservation.

An opportunity has arisen for me to visit New York City, in October, the Fall. It's a smart time to visit, I realise, because autumn brings such spectacular colour to New England's trees and landscapes, and because so many birds visit the city at this time. They are only commuting, passing through, on the whole, but the numbers and variety are bewildering. And Central Park is the place to see them. I might even find a Goshawk there. North America shares this species with Eurasia, although it comes in a range of subspecies across the northern hemisphere. And although they aren't regular migrants, as such, they do have occasional 'irruptions' south in autumn and winter, I have read. And they do occasionally turn up in Central Park.

Reading up on all this in advance of the trip, I pick up this book from the shelves. Another charity shop find, from some while back. One I've only dipped into, always meaning to read in full. Now I have the pretext for reading it. Having started, I can't put it down. It's a small hardback, called *Red-tails in Love*, by Marie Winn. It is the story of Central Park's Red-tailed Hawks, which began with the hawk that arrived, in 1991, to live right in the middle of New York, in and around Central Park. The Red-tailed Hawk is a Goshawk-sized raptor, with some (but not all) of the rapaciousness of a Goshawk. After a few unsuccessful attempts, Pale Male, as that pioneer, city-living Red-tail became known, and his mate finally reared young in 1995.

The nest site the birds chose, of all places, was 12 floors up on a building on 5th Avenue, overlooking the treetops, ponds and playing fields of Central Park. Perhaps appropriately, they built their nest on the anti-pigeon spikes intended to protect a "bas relief depicting two sad-faced cherubs," as Marie Winn describes it. The nest site is opposite the apartment of, of all people, Woody Allen.

That first nest in 1995 coincided with the city's "best year in recent memory," as the *The New York Times* put it, as other visitors began to be tempted back to the city. The hawks have been studied by a dedicated band of followers ever since. They have inspired many urban dwellers to develop an interest in nature. Living where they do, and given their size and magnificence, not much that these hawks do escapes notice. Nearly 20 years on, Pale Male is still going strong, street-smart and park savvy, though his partners have changed, due to fatalities to traffic and secondary poisoning. Marie Winn is their historian and biographer. Her account of their dynasty in the park is compelling. I am eager to see the place for myself.

Putting the book down I am more impatient than ever about finally getting the chance to explore New York. Among its many attractions I want to meet these Red-tailed Hawks. I wonder if I might also meet Marie Winn. Why not try? I find her on the Internet, and get in touch.

She replies almost straight away, with details of her group. "We're a very informal bunch. Right now a small group meets every night to look at and try to identify large and beautiful moths attracted to a certain tree. People just show up there – nothing pre-arranged. Last fall and winter we were quite involved in following a screech-owl family. I've been steering people towards nocturnal events – because that happens to be what I'm writing about now. I'd be glad to steer you to a few group leaders who'll be doing organized walks then.

"P.S. Not much red-tail action in beginning October. That scene will certainly heat up by the beginning of February ... If the weather gets a little warmer there is a regular bunch of insect-watchers at Shakespeare Garden you might want to check out.

"There's a regular bird walk every Wednesday morning at 7am, meets year round – the same group I wrote about in *Red-tails in Love*. You are welcome to join us. Very low key. We meet at the Park entrance, 72nd St and Central Park West." I tell her I'll aim for the bird walk, hoping that Marie and her posse might be able to introduce me to her hawks. But as fate would have it, it is Lola – the resident female hawk – that I meet first.

Sunday

Early morning. The park is deserted because of rain. Not a real rain, barely enough to clean the streets of pigeon poo, but a rain nevertheless. Around midday, the cloud begins to thin and the 'dimmer switch' to be slowly raised. I emerge from under my umbrella. The park's birdlife has begun to stir. Despite the wet, the place is stiff with birds, as it tends to be during spring and autumn migration. Migrants in their thousands funnel in to this rectangular oasis amid the towering buildings of the city. Hundreds of species have been recorded, including raptors of many kinds, mainly passing overhead, but some stopping here to refuel. Goshawk included.

I hadn't really expected to meet the resident hawks – at least not without being led to them by a native. Outside of the breeding season, as this is, their whereabouts are less predictable. But it seems that no sooner has the rain relented and the sidewalk begun to steam, than hurtling towards me through the trees comes a hawk, a big one, a female, on half-folded wings. She clears my head by a matter of feet and sweeps up into a tree by the path, hunched. A Grey Squirrel in the next tree along twitches its tail and chitters. The hawk pauses only briefly before launching again out of the tree and continuing on her way, over the Bethesda Fountain and on to another oak. Follow me, she might as well be saying. So I follow. This is probably her first chance to hunt today. Like the squirrel, I can tell this raptor means business. She is hungry.

She settles among the foliage of the oak's outer branches, partly concealed, about 10 metres up: a vantage point for hunting. I settle myself down at the base of a nearby elm: my vantage point for hawk watching.

"What have you got?" asks a passer-by, following my gaze.

"It's the Red-tailed Hawk," I explain, and notice that the questioner, and the others now mustering around her, are carrying or wearing binoculars. A New York street gang of sorts, with their gang symbols.

"Are you the Regulars?" I ask them.

The Regulars is what Marie Winn in her book calls the band of birdwatchers who have been closely following the fortunes of the hawks. It seems that, if not *the* regulars, they are certainly regulars. They confirm that this bird is Lola, the resident female. This is her 'hood'. Passing migrant hawks are not tolerated by Lola and Pale Male in this part of the park. They will be encouraged to be on their way, to keep on moving south.

A short while later the birders move on to look for scarcer visitors. Lola, who arrived five years ago as a juvenile, is an old friend to them. I, meanwhile, am still getting acquainted. I look around. A lone Feral Pigeon is picking around in the grass beyond the footpath. A Grey Squirrel is foraging busily. Both are oblivious to Lola's gaze on their backs. She cranes sideways and backwards for a view through the oak leaves. I wonder if and when she will choose her moment. Small dogs trot past on reins. I can't help fearing a little for their safety too.

I look from the hawk, to the pigeon, to the hawk, to the squirrel, and back to the hawk – craning, blinking, occasionally gaping slightly. Both of us, I suspect. It's not often you get views like this of a wild raptor. Back home in the UK, near London, where I now live, there are no urban-dwelling *Buteo* raptors this big, this habituated to crowds. Despite the superficial similarities, our Buzzards are not as equipped as Red-tails for catching prey like adult pigeons.

Many urban areas, globally and not just in the UK, have issues with Feral Pigeons. Local authorities everywhere struggle to control their worst excesses. The Mayor of London earned notoriety for his campaign against them. Falconers have been hired to fly Harris Hawks (an American species) in Trafalgar Square to keep the pigeon flocks at bay. Animal welfare campaigners have been up in arms. In the absence of anything akin to Red-tails, our pigeon flocks get bigger, mangier and messier.

Only the Peregrine Falcon, now establishing quite happily in towns and cities like London, could be called a regular predator of Feral Pigeons. Peregrines prefer to keep themselves to themselves, like the ones that roost on the tower of the former power station that is now our Tate Modern building.

Meanwhile, back in Central Park, these and many more thoughts circulate as more passers-by pause to smile at and even in one case photograph the squirrel. I point out the poised hawk to a few walkers who are eying me curiously.

The tension is building, as is the hunger. Lola's I suspect, and my own. I remember the half-eaten breakfast of banana and nut muffin in my bag. I throw some crumbs to the pigeon – which walks round in small circles but doesn't eat – and the squirrel – which evidently prefers acorns. An hour or more passes. Another young couple ask me what I'm doing. I point out the Hawk. They are amazed – native New Yorkers but not yet tuned in to the hawk programme, but eager to know more. Just as I am explaining, right on cue, that's when the branches bounce, droplets fall and Lola swoops, noiseless, almost in slow motion. A group of people on the path notice the hawk immediately as she descends. The pigeon does not. Accelerating, Lola is on her quarry, toppling slightly sideways onto an 'elbow' of outstretched wing as she strikes, pale belly exposed above long, outstretched, feathered legs. The pigeon squirms in a straitjacket of talons. Gaping, Lola turns to face her gasping audience, only feet away. I sense no shock among the onlookers, just a kind of awed admiration.

* * *

But why had Lola waited so long to strike? I think she used these passers-by as a screen, behind which to make the attack, hidden from her quarry's view. She may not have needed to. There was something abnormal in the lethargy of this pigeon. It had sat there for as long as I had. I felt like I'd got to know it. We'd nearly shared elevenses, after all. The softer side of me felt a little remorse. The logical me knows that this isn't a bad way for a sick pigeon to go. It is actually, in every conceivable way, for the best. Mind you, one of Lola's predecessors died mid-meal while eating a poisoned pigeon on the Metropolitan Museum of Modern Art. I can only hope no such factor has been at work here.

Curtseying, Lola opens her great wings wide and lifts her catch up onto the bough of my elm. There, back-lit by high-definition, New England autumn sunlight, she begins to pluck her catch. Feathers drift down onto the path and onlookers below, and for a full half-hour the hawk's lunch is witnessed by a succession of mostly gobsmacked visitors, including one who takes my email address and later sends a photograph. I forgot to bring my own camera.

Meeting Marie Winn is no such matter of chance, although in the gathering dawn of the following day I have to jog the last few blocks of 72nd Street to make the appointment with her: 7am, Central Park West for weekly bird walk. She introduces me to some of the Regulars from her book, and some more of the birds of the park she knows so well.

She has written another book: *Central Park in the Dark*. It's about all the other creatures that live in this great, green lung of this awesome, square-canyoned city: crickets, beetles, bats, raccoons. She's helping the smaller critters to share the limelight with 'charismatic megafauna' like Lola. It will add to the new image of New York as a clean, healthy city, capable of nurturing a functioning ecosystem at its heart.

In researching the book, Marie has had to spend quite a few nocturnal hours among the park's woods and thickets. A petite figure, with an open, kind face and dark, inquisitive eyes, she carries a powerful little flashlight, to illuminate bugs and, she reckons, to fend off assailants, should the need ever arise. You get the sense that in today's – or even tonight's – Central Park that is about as unlikely to happen as anywhere else.

"You talkin' to me?" she jokes, brandishing the little torch with the big beam. I think Travis Bickle would have been proud.

Monday

I take a train up the coast to Boston, bowled over by the scale of the wildness, the yawning coastal wetlands and the procession of large birds of prey gliding over them – mainly Broad-winged Hawks, moving south. I look, of course, for the Goshawk, but cannot be sure I see one. Some follow this raptor passage in the Fall, moving to the southern states for better weather and better prospects of winter survival. They tend to move a little later than this, if they move at all.

The Goshawk in the US is the same species as Europe's – officially known as the Northern Goshawk. The only northern hemisphere goshawk, and found right around the northern half of the globe. But it is classified as different subspecies – *Accipiter gentilis atricapillus* and *Accipiter gentilis apache*. There are some differences in how they live, in the US, to those in Europe.

For one thing, US goshawks live at lower densities; that is, there are fewer of them, where they occur. They are also less productive, producing fewer young, on average. They are more restricted to forests, and less inclined to move out of there into more open country, or to nest in small woods and even more populated places. They are also more likely to hunt mammalian prey.

These differences in Goshawk ecology in the US probably reflect greater competition with other species, notably Great Horned Owls and Red-tailed Hawks. Goshawks in the USA are likely confined to a more specific niche. There are other closely related hawk species in the USA, whereas in Western Europe there is only the Sparrowhawk among the *Accipiter* hawks. There is also a little less size difference between the male and the female, and they tend to be darker and more likely to have the red eyes of the birds I first knew in that book I had as a child.

I will have to return to the States on another occasion, with more time and local guidance, to look for the Goshawk here.

Chapter 3

LOOKING FOR T. H. WHITE

Saturday

Returning to the British Isles by plane, you encounter it as the passage Goshawk might: a group of islands – two large, many small – on the edge of the vast land mass from which you have beaten and blown. Arriving at East Anglia, say, your choice of woodland pit-stop is limited to spinneys, in the main, with one or two blocks of substantial conifer tract, such as Thetford in Breckland. Cambridgeshire and its fens you might think almost bereft of options. Bedfordshire beyond becomes more promising, even offering a ridge off which the winds from the flatlands might generate a welcome updraft. Looking at the map of Europe, it seems feasible that the curve of the coastline sends migrant raptors and others towards the great doorway to southern England that is the Wash, the yawning maw of an estuary where the River Great Ouse meets the North Sea. From there, the logical thing would be to follow the river inland, with its ribbons of gravel extraction sites decorated with the waterfowl and gulls that abound on them.

Immigrant Goshawks have been noted coming in off the North Sea at watch-points on the east coast, and once or twice reported from offshore oil rigs and boats, where they have paused for breath, and a spot of hunting in high-tech surroundings. But mysteriously, only one foreign-ringed Goshawk has been recovered in these islands. A male bird, ringed as a nestling earlier that year in Norway, was caught by fellow ringers in Lincolnshire in 1994. They logged its leg ring number before releasing it.

In lunch breaks I have taken to scanning the shelves in the RSPB library, carefully sliding out likely looking texts, scouring indexes for traces of Goshawk. The basic activity of looking under 'G' in so many books tells its own story. It's partly displacement activity, I am sure, for finding the real thing. It's like splitting rocks for fossils. Most flake open and turn up nothing. The bird simply disappears from most of the literature produced over about an 80-year period, from the late 19th century; unless, that is, the book is at least in part about history, or extinctions. Otherwise, across several generations the Goshawk is gone, mostly forgotten.

Books of British birds produced up to the 1970s often make no reference to it whatsoever. Others contain a few lines, a footnote, about the bird that once was with us, but by now is vanished. Most could not describe the Goshawk among the creatures encountered in farm, field, hill, dale, mountain and forest. There weren't any resident to encounter, and you sense that few of the authors really know this bird at all.

Over a period of weeks I have leafed through dozens of these books about living with and appreciating nature, from the turn of the 20th century. The style evolves, from the lyrical and personal to the spare and factual. With the help of better and steadily clearer photography an increasingly urbanised, wider book-buying public was encouraged as the decades passed to take an increased but still mostly passive interest in natural history. And with these changes in representations of nature, over time it was becoming steadily less necessary – or acceptable – to take wildlife home: whether pinned, mounted, worn or merely stuffed.

In more recent books, the mentions I am seeking are brief and follow a familiar theme and pattern. The Goshawk, they are fond of repeating, is a bird of remote forest, hard to see, tricky to census, doesn't really want to come out of the woods, population probably growing but still officially the same as it was, no point really in getting too excited about this bird of dubious origin, with its barely legitimate status: may not even really be supposed to be here, is the subtext I detect.

"We've always been a bit iffy about the Goshawk," someone prominent in conservation said to me at a conference. I thought it significant. It's not the only time I've heard the word 'iffy' used in connection with this hawk.

Older books, from the earlier 19th century, I am pleased to say, have more. I photocopy the entries. I realise what it is that I am doing here: I am trying to fill a Goshawk-shaped hole in my world by gathering small fragments of Goshawk knowledge around me. I am trying to see representations of it to compensate for what has begun to feel like the impossibility of real, first-hand experience. Above all, I am trying to piece the bird back together, breathe new life into it.

In that largely barren, early to mid 20th century period when the Gos was absent, there was one book written on the subject. Perhaps unsurprisingly it isn't a study of the bird in its wild state, by an ornithologist. It is about a captive bird, written by a former teacher of English literature, a medieval historian, a novelist. This is the first and best known of the books dedicated to the species. There is one other, published here a few years ago; a detailed and very impressive scientific monograph by Robert Kenward.

The Goshawk has long lurked in the thicket of book spines that line the horizons of the study at the back of the house here. It was written by Terence Hanbury White. T. H. to his readers; Tim to his friends. I picked it up in an antique bookshop in the old town at the summit of Stirling's cobbled lanes, when I was a student there. I know I read at least part of it while living in Bridge of Allan, at the base of tree-clad slopes, and imagined myself living as White had imagined he could, having "reverted to a feral state". The definition of the word feral may have changed since White's time. For him it had romantic associations with being "ferocious and free," while for us now it connotes semi-tameness, or part-dependence on human society; more like city pigeons or farmyard cats than the alpha-predator of our deepest forests.

I have taken it down once again from the shelf. The cover has a painting of a Goshawk on a bow perch, head drooping, ankles strapped. It is a study in misery, defeat. I have discovered a more recent version, published by the *New York Review of Books*. It has a much more suitably dramatic cover: the bird in action, subliminally. And the preface has been contributed by none other than my NYC friend Marie Winn. It has a rare photograph of T. H. White too. With his trim moustache and sleeked hair he looks a little like Errol Flynn – handsome, debonair, dashing. He was said in later life, bearded and heavier, to bear a striking resemblance to Ernest Hemingway.

The Goshawk was first published in 1951, although it is a journal of events kept as long before that as 1937. Dissatisfied with it at that time, he stuck the manuscript in a drawer and tried to forget about it, while he wrote other things, most notably *The Sword in the Stone*, first in the quartet of novels based on the legends of King Arthur which would make him famous, and quite a bit of money. It was selected as Book of the Month in the USA, guaranteeing huge sales and financial security of a kind for an author at that time of uncertain means.

The Goshawk is a dense thicket of prose, a tangle of candidly shared emotions. White's passion is clear but he was not yet an expert on either birds in general or falconry in particular, and this was the basis for his own ultimate misgivings about releasing the manuscript. But he was determined to learn and his enthusiasm illuminates every page. The book describes White's relationship with a captive Goshawk, written as a diary over a period of weeks. He was prompted to purchase the bird after reading about the ancient art of falconry in medieval texts, in which he was immersed as a student, and as a teacher and author on books linked to the period. He was amused by country sports, and had just written a satirical account of them in a book he called *England Have My Bones*.

When reading up on medieval falconry he came across "a sentence which suddenly struck fire from my mind," he explained. "The sentence was: 'she reverted to a feral state.' A longing came to my mind that I should be able to do this myself ... To revert to a feral state! I took a farm-labourer's cottage and wrote to Germany for a goshawk."

White tried to man – or train – the bird using medieval methods, in part because he was interested in that period of history, but also, he claimed later, because he wasn't aware of any other way. The medieval way involves keeping the hawk awake until it gives in to fatigue and submits to sleeping on its captor's fist. The austringer (as the shortwing hawk falconers were and are still more properly known) would usually cheat – using back-up from others to take shifts while he or she rested. White took on his Gos (and that is what he called his hawk, a male) single-handed, again in part because he was living alone by now, in this cottage in the woods, and because he relished this kind of challenge. It might also provide him with a more interesting narrative to recount. It has been described as a love story, or a story of attempted seduction, and it has all of that intensity.

It is impossible not to be seduced by this in places harrowing account of his love-hate relationship with the intransigent raptor, a saga strewn with error, frustration and mishap. Behind the wit, intelligence and erudition of White's prose there is a strong hint of tragedy, of a man at odds with himself and humanity – the grown-up part of it, at any rate. He is disdainful of the school from which you sense he feels he has been lucky to escape, and he makes bleak references to the impending war in Europe and what its approach reveals about the failings of our species.

Rather than dwell on the lack of insights on wild-state Goshawks, I have become intrigued by the author. Despite sharing his fixation with the bird, I'm not sure I have fully understood seeking freedom and wildness by attempting to deny these things to the bird that epitomises them perhaps better than any other. He does note the apparent contradiction: "I had only just escaped from humanity, and the poor gos had only just been caught by it."

Friday

Halloween night. The back garden Eucalyptus seethes and writhes, clinging fast to its leaves, dark and sinewy against a lighter sky, shaking its bunches of leaves on oil-filled limbs like a manic Antipodean dancer. The wind is mild, refreshing to stand out in, even at midnight, as the gusts surge stronger. Rain flecks and leaves scatter on the wind, and drip into the already brimming water butts, and into the ponds.

Saturday

Morning brings some respite. Gulls and crows tumble in the residual gusts, bowing toward earth from their loose, windborne flotillas. They are no doubt sensing worm-lined puddles in the ploughed fields and pastures around.

I'm reading Tim White's biography, by Sylvia Townsend Warner. He was born in India at the turn of the 20th century, soon traumatised by violent quarrels between his parents and their humiliatingly public break-up. He was shipped to England to be educated, sent to Cheltenham College, a prestigious private school close to England's western border with Wales. I drove past it recently, an elegant old building, faintly redolent of a Victorian prison. I am sure it doesn't happen nowadays but it is recorded that White was brutalised there. It helped him develop his need to work, learn and write, to free himself from his unhappiness.

He graduated from Cambridge, following a thesis on Malory's *Morte D'Arthur*, although he claimed not to have actually read it while a student. He took a teaching job at a rural private school in central England, among some of Europe's finest landscaped gardens, created in the 18th century, with more than 40 monuments, temples and follies. After a few years of this he wrote to his friend David Garnett:

"It is time to face the issue. I hope to get out ... and live in Scotland on £200 a year. I want to get married too, and escape from all this piddling homosexuality and fear and unreality."

White's Arthurian legend novels were later compiled as *The Once and Future King*, in turn condensed into the Broadway musical *Camelot*, and later the hugely popular Disney films. J. K. Rowling cites the character of Wart, the young King Arthur, as the spiritual forebear of Harry Potter. It is plain that White's influence has been widely felt, yet he remains curiously marginalised in literary history.

I'm not sure why it has become so important to me, but I long to know where this man and Goshawk battle of wills took place. It becomes the next step in the search for the Goshawk, even though it might not bring me any closer to the actual bird. There probably hasn't been a Goshawk in those woods since White flew his captive bird there back in 1937. I can't even be sure the woods are still there.

Sunday

It turns out that this cottage, though remote, was in a neighbouring county to mine. All through reading the book I hadn't imagined it was so close to here, although to be fair White gives few clues. The cottage seems to have been a few miles from the school, which I have worked out is about an hour from here.

By late morning the winds have tailed off completely and sunlight is bouncing off the fields and surface water of the roads, not strong enough to evaporate them but bright enough to dazzle and heighten the spirits. I set off for the exploration, not sure what I hope to find.

The occasional shotgun report breaks the calm, from afar. Gulls crisscross the sky like skaters on a frozen lake. A Kingfisher zooms through the park, between the many ponds. Geese graze the banks of the Elysian Fields. The school is part-wrapped in scaffold and tarpaulin. There is a church buried in evergreen trees. It feels like a place where the pupils might hang out at breaks; darkened rooms within the outside world. Thrushes and pigeons scatter overhead. Woodpigeons, Redwings and Blackbirds are spurting out of the dense foliage of Yews and Cypresses at my approach, flashing pale as they catch the sinking sunlight. This could even be prime Goshawk terrain, but I see only Kestrels today.

I try the National Trust visitor centre. The assistants are very helpful, but have no idea who or what I'm talking about. I wander round the gardens and follies of the estate, which were handed over by the school to the Trust to manage in 1989. There are fountains that make music from their drips, there is a parade of worthies, including one bust which looks like the photograph of White on my book but turns out, on closer inspection, to be Sir Francis Drake.

A feature called The Sleeping Wood has been put back together, an example of a woodland garden, the perfecting of nature, much in vogue at one time. There are waterfalls, archways, grebes, Coots and swans on elegantly sculpted little lakes with tree-covered islets, long vistas to Hadrian's archway on the skyline that dominates the whole, an avenue sweeping down and up again from the front of the school for all of two miles.

Monday

I refer again to the biography. The cottage was evidently a gamekeeper's, in a wood, half a mile from the nearest road, with a barn and a nearby fishpond.

Maybe Marie Winn has some clues. "I have been reading T. H. White's biography with avid interest I tell her. I was even prompted yesterday to try to find his cottage. Didn't have time to see the exploration through but wondering if it's still there somewhere. I found there no immediate trace of it."

"Oh Conor, you've struck on a subject very dear to my heart," Marie replies. "I fell in love with T. H. White's writings in my early teen years and never really fell out of love. For years I treasured my rare, out-of-print copy of *The Gos-*

hawk, and when invited to write an essay for the new edition I quickly accepted. I'm really sad to think that the gamekeeper's cottage might no longer exist."

***

I have contacted the school, and ordered a map online. In the meantime I would also look out for an old map. Maybe the landscape differed in the 1930s. In any case I would have to wait for another opportunity to visit, and to search.

Friday

The pavement in my street is damp, and there is a hint of dampening drizzle in the air. Fallen autumn leaves are beginning to line the pavements. There is a buzz of activity among the Rooks and Jackdaws at Esme wood. Some are airborne, wheeling, hanging around. Many are gathered on the topmost, already bare branches of the trees, calling – the Rooks gruff and hoarse, the Jackdaws higher pitched, almost sneezing. A raptor comes in to view from beyond the treetops, gliding slowly, The crows take off, noisily. I take it at first to be a Buzzard, and some of the Rooks rise, as though to escort it, as would be the norm with a Buzzard. A couple of Jackdaws dive at it from above as it passes high over the road. It snaps at one of its assailants. On the whole, though, despite their overwhelming numerical superiority, the other birds are avoiding it, and watching it carefully.

This bird looks longer tailed and a little more sure of itself than a Buzzard, less rounded and moth-like, more streamlined. There are always Buzzards around now, but they don't usually look like this, and there is something about the time of day, the attitude, and the weather, that doesn't quite fit. It just looks much more ballsy than a Buzzard. The vibe around it, among the other birds, is different. Respect mixed with diffidence. It sails through and is gone. I turn the image of it over in my mind. What would a Goshawk look like up there, I ask myself in the immediate aftermath. Can I rule it out?

Goshawks move around in the autumn. They can, in theory, turn up anywhere. Young birds move away from their birthplaces and try to establish themselves in the world. They will tend to be drawn to sites that resemble where they've come from, if such options are available; somewhere they have a bit of peace from disturbance, by anything – human or otherwise – that might be hostile to their presence. Somewhere that provides food to eat. Not necessarily in that order. A bird's first priority is to eat. If it can't do that in what it perceives as ideal habitat, it will either settle for a less ideal habitat, or starve.

It was autumn when T. H. White found himself in need of a second Goshawk, his first having absconded. His only option, by this time, was to purchase a passage bird – one taken not from the nest, but from the air. A bird that had had an initial, independent existence, and taught itself how to hunt, and to fear people. There is a place in the Netherlands called Valkenswaard – translating as 'falcon's field' – one of the centres for hawk trapping in Europe. It is a ridge in otherwise flat country, raised ground followed by migrant raptors, a bit like our greensand ridge between my village and The Lodge. To catch the birds, trappers deployed shrikes, renowned for their abilities as hawk-spotters. These would give the signal when a hawk was in view – long before the trappers could see it. A tethered pigeon would then be set in place on a perch, to lure the hawk in. The trappers concealed themselves in pits behind it. As the raptor drew near, the pigeon would be withdrawn. Another pigeon, on the ground, would draw the hawk closer in, and a clap net would do the rest. The practice is long since discontinued. European falconers must now breed their own stock from already captive birds.

What do other falconers think? "Even now, after training so many, their initial wildness still surprises me," falconer Phillip Glasier once wrote. "As does the swiftness with which their whole attitude suddenly changes towards humans once they learn that no harm is meant."

Writing about his working ferrets, he said: "We kept them in a hutch opposite her perch on purpose to get her accustomed to them. If you want to stop a hawk going for any particular thing, this is the way to do it. They soon get used to chickens, for instance, yet still fly pheasants unhesitatingly."

He also describes visiting the royal Sandringham estate the time he found himself "faced with the problem of greeting the Queen in a dignified manner with an irate goshawk bating off my fist."

I admire those who have the time and skill to fly captive hawks well, and legally, and do justice to the birds. I prefer my raptors wild, even if it means I don't get close to them very often. But I know that one – and probably the only – way of guaranteeing a good view of a Goshawk in the UK any time soon would be to find one in a falconry centre. I hit on this apparently fail-safe scheme while in correspondence with Jemima Parry-Jones, MBE. Mima runs the International Bird of Prey Centre in Gloucestershire, near the border with Wales, and is probably as renowned a falconer and captive bird of prey expert as there is, maintaining the family tradition. She has provided invaluable insights to conservationists on aspects of aviculture. It is sometimes necessary of course to bring birds into captivity as part of efforts to save them in the wild.

I ask Mima if she has a Goshawk among her extensive collection of raptors. She replies that, as it happens, she does have one right now. She is looking after it for someone else. It's a male. This gives me another pretext for heading due west across England, to that oddly inaccessible and off-the-beaten-track part of the country that is inviting in so many other ways. In addition to visiting Mima, the centre, and the Gos, I could take in Hay-on-Wye, the second-hand-book capital of the country. There would also be a chance to drop in at the Forest of Dean, where wild Goshawks are known to lurk.

Sunday

I head west in the morning, via Milton Keynes and Buckingham to Oxford, taking the A1 south to the orbital round London, to the motorway that fans off that race track towards Oxford. This is Red Kite country. By the time I turn off the motorway about half an hour later, I have counted 29 of these spirit-lifting, languid, angular-winged birds, drifting weightlessly on the autumn breeze, against a blue sky and hazy clouds, glittering autumn beech woods, the colours of which are mirrored in the auburn, buff and chocolate brown patchwork of the birds' undersides. They are like little badges of autumn pinned to the sky, stringless stunt-meisters wobbling on the wind.

Kites are so good at being airborne that it doesn't cost them much in terms of effort to be so. They might as well be on the wing, as perched in a tree. They are relatively social, as birds of prey go, and tolerant of people. They also like scavenging, and they tend to be very attached to where they came from, hence the density hereabouts. This is close to the epicentre of the Chilterns reintroduction programme that the RSPB and partners began about 15 years ago. This project, in terms of numbers of Red Kites, has been a raging success. So much so that it has been attempted in about ten other places in the UK. Some have even used young birds from these Chilterns nests. The Red Kite is a really obvious symbol of conservation success, and of our landscape's capacity to welcome back things that we have lost, or driven out. They are highly visible, which helps. The contrast with the Goshawk's situation and habits could hardly be more stark.

And, lovely and uplifting as 29 Red Kites undoubtedly are, it is to the Goshawk that I am headed. I get snarled up in Cheltenham, which allows a long look at T. H. White's old school. Knowing what I now know about his unhappy experience there it takes on some of the foreboding of a Victorian prison. Then it's to Hereford and then the last leg to Hay-on-Wye, which lies on the scenic, snaking river that forms part of the border between England and Wales. After Hereford the roads are mercifully quiet. The fifth most populous large country in the world seems suddenly distant.

I pick up a leaflet listing the 30 or more second-hand bookshops here. I am soon looking for the Goshawk in print. There are a few old books I know would be useful that I don't yet have, and I am hoping I might turn them up here. I find a reasonably recent bird report for Herefordshire. It says there were reports from 25 well dispersed parts of the county, in 2004. The Goshawk is clearly out there.

I don't find the titles I am looking for, but I do find a few useful leads. In fact I do more research and note-taking than actual buying, from antiquarian titles that are out of my price range, but which reward the customary quick look in the index for Goshawk, sometimes leading to a nugget or two.

Bannerman notes a curious absence of Goshawk from Welsh history: "There is no record for Wales, even of a casual visitor." He also makes some interesting observations on the ethereal quality of the Goshawk's plumage: "As H. L. Meyer noticed over a century ago, the beauty of a goshawk's plumage is enhanced in life by 'a kind of bloom' which ... fades soon after death". Others have described this "evanescent bloom of ash-colour on the living bird, which fades away shortly after it is dead". Austringers have long used an eastern term – in *yarak* – to describe a hawk in this primed, glowing condition, when it is in the peak of condition, hyper-alive, and just right for hunting. It barely translates second-hand.

A combination of large tractors and narrow lanes means I don't get to the bird of prey centre until just after the first falconry demonstration of the day has started. After that I am free to explore the birds in their pens and cages. I pace them out methodically, enjoying the birds in the main but eager to get to the male Goshawk. I am still trying to consolidate my understanding of what this bird is and looks like, to get a sense of it properly coalesced in my mind, from the real thing. The books can only teach you so much.

I am running out of raptor cages to check, and still no sign of the Gos. I find Sparrowhawks, so I must be getting warm. A couple of males are sharing one pen. They are being rehabilitated for later release. No Goshawk in the next cage, or any of the others in the row. Hobbies, Peregrine, Eleonora's Falcons from the Mediterranean. But no Goshawk. I have one last block to cover, composed of larger aviaries, next to the restaurant. There are four massive Griffon Vultures in one room, then eagles of different kinds, and more vultures. I find African Fish Eagles round the back, and Tawny Eagles.

I am down to the last aviary, the one at the end of the row facing the trees at the back of the Centre. I pass Buzzards, White-tailed and Stellar's Sea-eagles. The last named point their necks to the roof and guffaw as I pass slowly by – the sound of Africa transporting me back to Kenya, issuing to the woods and fields of Gloucestershire. I am giving up hope of there being a Gos. Even here, I am failing to find. The final cage, and literally the last bird, is now in front of me. There are three different species in this cage. One is a small eagle – I guess Tawny Eagle or maybe a large buteo of some kind. The second is a small white vulture.

The third and last bird is sitting high on a perch in a corner. It has its back turned but its head twisted round, glaring back over its shoulder, its wide yellow eye with beady black pupil redolent of shocked indignation, fear and anger. It is more than a little disconcerting. The eye is large in an already proportionately outsize head. The back is blue-grey and the tail banded with darker bars. It is almost like a thicker-set Gos, but surely not an actual Gos. Or is it? I have run out of options. It is motionless. There is something different about it to all the other birds I've seen – a Gos madness in its eye, a penetration from its stare, the glare that epitomises the hypnotic quality of raptors, and of the Gos in particular. It reminds me of the thing that got me hooked at such an early age, the pictures in that book, the Hamlyn guide. But is it a Gos? The bird is tucked in the far top corner of this cage, but I feel sure, despite willing it to be, that it can't be a male Goshawk – much too large, surely? And the proportions just seem wrong for a female Gos. Plus I know enough by now to know that you wouldn't put a Gos in a shared cage. With anything. I have to be sure. There is no labelling on the cage, so there is no way of being sure without asking.

I finally track Mima down in the cafeteria, just next door. I introduce myself – we've never met although we have corresponded – and I ask what the birds are in the mixed cage. "Tawny Eagle, Palm Vulture, and Indian Snake Eagle," she says, without pause. I ask her where the Goshawk is, thinking it mustn't be on show, that it is back-stage. Perhaps she'll offer me some kind of peek at it, on its terms. "It's not here any more, it had to go back," she reveals. She told me later that she would always suggest to any visitor that if they are particularly wanting to see a specific bird they just need to make a definite arrangement beforehand.

Disappointing, of course. But the other birds, and the snake eagle especially, have been some consolation. That manic, fool's-Gos eye haunts me for the rest of the visit.

***

Saturday

Home again. I drive south-west into the 'sunlit uplands' of the county, to a picture book village in the part of the county where the roads start to climb on gentle slopes, to rolling hills. It lures me now, as it feels to me like the place from which the Goshawk would likely launch any bid to repossess my neighbourhood. If I actually have seen a Goshawk near home, or found evidence of one, I would guess it somehow links to here. And the Goshawks that others may or may not have seen could have had their origins in this part of the county.

Saturday

It being autumn, it is also a good time to be in woodland,.perhaps particularly so on a wet day, with the leaves starting to shed themselves in steady drops, and occasional squalling flurries. My wife Sara and I walk the narrow road that is its only street past thatched cottages with ornate timbers and leaded windows, and enter the estate via a gate. A sign asks walkers to keep their dogs on a lead, and offers the incentive that there is rodent poison to consider. Secondary poisoning is a threat not just to domestic pets, but to birds of prey. A rat laced with anti-coagulant is an easy and lethal meal.

A Muntjac Deer slopes off, hog-like, into a field that is framed by tall oaks and pines, which in turn edge a conifer plantation. This is Buzzard and even Red Kite country now, and Ravens are occasionally reported from here too. I think I might have heard a distant croak. A Hornet sends ripples from the stream below a bridge, as it sails in pointless circles on its back, having tipped in while having a drink. I rescue it on a long branch.

I pop my head into an open wooden barn, to check for owls, or their pellets. I see a Jackdaw's old stick nest, and I hear a movement in the rafters above. A Little Owl. It bounds through the open loft, and out of the building, round on broad little wings and up in the oaks of the wood.

Dark feathers are visible in the roadside grass, and I wonder at first if I have found a dead Raven. Its bill looks thickset and heavy, its throat grizzled. It has barely a mark on it, and I worry that this could be a victim of secondary poisoning, and be itself a possible threat to a further scavenger. I photograph it with my boot beside it for scale. It looks suddenly huge, but it can only be about 18 inches long – crow-sized, like a male Gos. Ravens are two feet or more long, their wingspan four feet across, big as a female Gos.

We reach a quiet cemetery. The first grave I notice as we pass is a Faulkner, Florence Elizabeth, who died in January 1941 aged 65, 'Peace after pain,' it reads. Close by, and under a huge, wide branching white pine, lie scattered the feathers of pigeon and crow, blended somewhat with the grass cuttings. The sun is now fully out, with wide stretches of blue between towering clouds, and glowering patches of dark cloud providing the ideal backdrop for glowing stands of poplar and bronzed beeches. A raptor cries unseen from somewhere in the midst of the thicket.

At this moment a hawk shoots overhead, about 10 feet only above our heads, having passed from behind. It veers and wobbles as it notices us below. And involuntarily I am saying, or rather warbling, *"woaahhhh wooahhh wooahhh"* – three-syllabled like an owl – because this is quite a big hawk, silvery grey below, and noticeably rounded in the breast. Though short-sighted, Sara has had as much of an eyeful of it as I have. "That looked like it might have been a Goshawk," I am saying. It was only visible to us for about three seconds, but it was close, and the light was good, and it was paler than I've ever noticed a Sparrowhawk to be at close quarters. Perhaps the sun was glaring off it. But it also looked chunky. It was in zoom mode, not spread-winged. The wings were flicking at its side, held in close for aerodynamism. Propulsion. It was rolling. Tilting. It was definitely a hawk. But which: female Sparrowhawk, or male Gos?

It is gone. We discuss it all the way back to the car, rattled. It confirms for me how difficult it is to be sure, and how easy it would be for male Gos, at least, to be being overlooked. No one would accept a record from me on this basis, not least because I wouldn't submit one. I could merely tell a few confidants that I might have seen a Gos here, and give a description. What I could also do is come back, and continue to look, and to try to find more evidence. I could come back again in the early spring, look for displaying birds. But I'd be confined to public paths. Perhaps it is better just to leave well alone.

There are other, minor, incidents, perhaps tiny clues, adding up. A short distance further on, at the other side of a field of long, untended grass, a terrific squawking erupts, and a Green Woodpecker explodes from the ground somewhere near the base of the row of Lodgepole Pines that lines the woodland there. It labours across this meadow in exaggerated rollercoaster bounds, deeply freaked by something. Jackdaws scatter in small groups, at canopy height. But I can't make out what might be disturbing them.

A sudden shower comes on and we shelter. It passes quickly. A very large dead rat lies by the road edge, the contents of its own flattened head being apparently vomited by its equally flattened mouth. I wonder if it's another secondary poisoning incident waiting to happen.

Back at home I delve in the identification guides once again, to see if I can glean anything I've forgotten, to aid the case for it being Gos v Spar. This only adds to my uncertainty. There is size variation even with female Spar and male Gos. The difference in size between one and the other could be as little as a couple of inches. But my strong sense from this bird is not so much its large size overall, but its chunkiness, and its silvery greyness. I think we might – just might – have seen a Goshawk today. Perhaps if and when I get to know this bird better, I will be able to reflect again and form a more distinct opinion. And I know I will be back with renewed optimism, to search for further tantalising glimpses, to enjoy the woods, from the required distance.

Friday

I have been looking again at photographs of the dead crow and shown them to another colleague. I notice something I missed at the time of finding this stricken corvid. It looks like wounding on the lower back. Missing feathers and bare flesh – perhaps the signs of where it had been raked by a hawk? That might explain how it came to die here, by the road. The hawk may have been flushed from the kill by a passing car, say. The feathering of the neck of the crow was very fluffed up, as though the bird had died in some kind of struggle. I show it to one or two other colleagues, and send the photos to an expert. He doesn't think it looks like a hawk victim. He points out no sign of plucking, which usually starts straight away. I hadn't noticed any stray feathers but then I perhaps hadn't looked closely enough.

November

Normal autumnal service resumes as heavy rain, high winds and flooding conspire to bring a balmy autumn crashing to a rude end. Severe weather warnings have been issued, as places ship as much as two weeks' worth of rain in a few hours. There is real rain on the window in the morning, the sound of childhood in Ayrshire. I watch the drops race each other down the glass, coalescing then snaking as they go. The wind is gusting, and the last of the ash leaves has left the tree. Gulls still sail overhead, and wagtails, perhaps enjoying the wash. Trees seethe in the mild wind as much of the autumn colour is dashed to the ground.

My wife Sara's mum Deirdre has brought a present – *The Peregrine* by J. A. Baker. It was published in the year that I was born, and, like *The Goshawk*, is widely regarded as a classic. Unlike White, however, Baker was writing about wild birds. And his account of following wintering Peregrines in Suffolk/Essex covers the darker months, from autumn to spring. I find a Goshawk in there. It is on the wrist of a hawker. Elsewhere in the book is a description of a Peregrine chasing a Sparrowhawk through a wood. It's not typical Peregrine behaviour to pursue in woodland ...

Baker revealed very little about himself in this book, but it has emerged that he had been diagnosed with a debilitating illness some time before he embarked on his study of the birds. It is as if he wished to live and see life and the world through them, to surrender his own identity to that of the falcons.

DECEMBER

A bitterly cold and fogbound morning. I board a bus in a neighbouring village for the journey to Oxford, for a symposium that brings artists and conservationists together to discuss new and creative ways to save the world. Oxfordshire is a well-wooded county, but there were just two reports of Goshawk in its latest annual bird report.

Event organisers Mark Cocker and John Fanshawe talk about their further researches into J. A. Baker. This is now written up in a fascinating preface to a revised edition of Baker's two books, *The Peregrine* and *The Hill of Summer*. On her death, Baker's widow Doreen released his notes and diaries, which have allowed new insights into this intriguing figure, about whom so little was widely known until now.

Between sessions I chat to artist Carry Ackroyd about our shared interest in the poet John Clare. The subject of Goshawks comes up. Not that John Clare would have known the species, long gone from fenland by the time he was lamenting the landscape's enclosure in the early 19th century. Carry tells me she saw one once, when she worked at what was then the Institute for

Terrestrial Ecology at Monks Wood, not far from me. At least she thought she had seen one. When she mentioned it to one of the professors there, he told her he'd also seen one, displaying. Corroboration helps. Hers had a highly visible leg ring on it – red, as she recalls. I quizzed her further about this, and it turns out she'd been really close to this bird, in the wood. She'd stopped to regard it. Her description is of an unusually large bird of prey, and she was struck at the time by how close it had allowed her to be. I wonder if the leg ring might in fact have been jesses – leather ankle straps fitted to captive birds. She says she can't rule that out.

It's an impressive line-up of speakers. Helen MacDonald gives a talk linked to her engrossing book *Falcon*, about the cultural importance of these birds. Afterwards I chat to her about her involvement in falconry, and her Goshawk, called Mabel. She very kindly gives me a CD of her Radio 4 series about her experiences training and flying the bird. It isn't all I speak to fellow delegates about, I maintain, but we get chatting to a man called Richard Hines, from Sheffield, also about the Goshawk, and specifically about T. H. White's book.

"It changed my life," he tells me. We talk about the cottage where it all happened. Richard shares my interest in knowing if it is still there. He's been working on his own memoir of former days with hawks. He doesn't fly them now, but his love for them shines through from our brief initial conversation. I have warmed to this man, sensing a kindred spirit. We swap contact details, thinking we might one day make a pilgrimage to 'Goshawk' Cottage, or at least the site where it stood.

Sunday

I have maps now – a new one, and one from the 1940s. The rains have returned, with a slight thawing, and sunlight is bouncing off the puddles again. I pack my mountain bike in the car to resume the search for the cottage. The car judders all the way on the first ten miles, shaking off the damp of idleness. The risk of breaking down makes me consider turning back but this is my last chance to search before the Christmas break.

On the drive there I turn over in my mind what I'm doing, what White was doing. He was in a sense looking for himself, and I don't think it gives too much away to reveal that he spends a large part of what is quite a short narrative looking for his Goshawk which, perhaps inevitably given his inexperience and outmoded methods, he lost. Perhaps his Goshawk is still out there somewhere, through its descendants. It is technically possible, though highly unlikely. Related or no, I love the idea of finding a Goshawk wild there more than 70 years on.

Of the nearby locations White describes in his book, I have found Three Parks Wood on the map. I work out from clues in the text that it must be west of the school. On the map it has no obvious buildings in it, it is quite small, and there is a footpath past it. The wood is also on the 1940s map, the same size and shape. Somewhere I can visit, at least.

I bump into a man walking up the track towards it, as I get off the bike to check the map again. I ask him about it, and he confirms that I've got the right place. He is off to feed the Pheasants there. He's quite interested in the White story, when I explain my interest, although he hasn't heard of the book and has no clues as to the whereabouts of the cottage. He points me towards an isolated house in the near distance. "It could be that one," he suggests. "You should go and speak to the man who lives there."

I trundle over there, and the man in question eyes me a little suspiciously from near the house as his young son comes over to ask me who I am and what I'm after, at the gate. "I'm looking for a house," I call over to his dad. He walks over, we chat. He's never heard of the link to any famous authors, but we speak at length about what he knows of the history of his house. We more or less conclude that this may well be it, unless it was one of the other places he has searched through in his memory, including some long gone. He – Clive – takes my details and says he'll get in touch if anything else occurs to him. I head off again to explore a few of the other options he's suggested.

I roll past Three Parks Wood, and peer in over the old gate. This is where White spent the night, lying under a grass-covered blanket, in place before the dawn to try to catch a Sparrowhawk. This was not at that time a protected species, although the methods White employed, using live baits, were a little dubious even by the more relaxed welfare standards of the age. His own welfare was compromised by some of these ill-starred commando manoeuvres, and the only thing he nearly caught was a poacher, who trotted past within ankle-grabbing distance. Today, there is nothing in here but a derelict caravan. Although decades have passed since White's day, and night, the wood still doesn't look mature enough for a Goshawk.

I pedal on, avoiding the deepest of the puddles, reaching tiny clusters of houses, chatting in places to locals that I meet. No one has any idea where this cottage might be, but there are one or two suggestions of people who might know.

Towards the end of the afternoon I am in a field, peering at the foundations of a former house, to which I have been directed by the last local I met. There is an old spring nearby. White's house had a well. I wonder if this is it. My phone rings. It's Clive again. "I've got someone I think you should meet," he tells me, sounding a wee bit excited himself.

There isn't a lot of daylight left. I have to return to the car, load the bike, and drive to a rendezvous point on the main road. From there I tail Clive to another farm, and into a yard. I follow him, stepping carefully through a small herd of cats and into an old farmhouse kitchen. At the table is sitting an elderly man who must be in his eighties, still dressed for work, beside a dusty old cast-iron range, with cats peering from the entry points to all its cavities. The man has his arm on the table, and is regarding me with studied interest.

After a short while and pleasantries I think he has decided that I'm okay. He has revealed that he knew T. H. White personally, and takes me to show me what he thinks are initials carved by the author on a door out the back. But this isn't White's cottage.

For that, I have first to go and meet someone else.

We drive out again and down the hill for a couple of miles and into another isolated yard. By now nightfall is complete. I am taken into a brightly lit, warm, large, cosily cluttered kitchen, to meet a man called Steve. My purpose is explained, again. Steve doesn't need all the details. He knows the author, and is clearly delighted that someone is this interested. He is soon producing his collection of books by White, and in no time is pressing some of these on me to borrow, along with a very welcome mug of tea and a biscuit. He has relatives who have cameo roles in *The Goshawk*.

"I love all this!" he beams, proud of his family's links to the author. We chat for a while about White's writing, about the school, about his family. But what about the cottage? This, it seems, is a little bit trickier. The cottage is still there, but it's too late to go now, and there's a lady – his aunt – who still lives there, and is protective of her privacy, and perhaps of White himself. She knew Tim when she was a girl. White was fond of children and his escapades and fads were an endless source of amusement to them. It's clear that I won't be able to visit this evening, but I might in future. Steve will let me know, but he's managing my expectations. He draws me a map, so I can at least, on another day, pass by at a discreet distance, and view from afar.

I leave elated, mainly to have met such a friendly and welcoming community and extended family in the heart of rural England, to have been received so warmly into their homes, and helped in what many might think a faintly eccentric mission. I love that the cottage survives. And people who knew Tim. I am smiling most of the way home. This is at least as fulfilling as finding the Goshawk itself.

It would be early spring before I can find a spare day to head over that way, to see the cottage. I don't hear from Steve again, and I want to leave it in his court. Meantime, I post Clive a copy of *The Sword in the Stone* for his children, by way of thanks for his help.

Sunday

After decades of mild winters, we are having a proper freeze this year. Mother Nature's counter-punch against the general onslaught of our warming effects. The ongoing spell of arctic weather is a shock to the system of wild birds. They need more food, there's less of it and it's harder to find. People care about this. A lot of bird food is sold. Most starving birds die unnoticed, unfound even by scavengers like Buzzards, which are also being found, dead or dying. My instinct is that the Goshawk would be quite handy and not short of easy prey, in these conditions.

Some other birds are making light of the sub-zero conditions. From the end of the back garden I watch a Kestrel as it evades the clumsy efforts of a Rook to dive-bomb it. After a minute or so of this I notice a Sparrowhawk circling some distance away, approaches the pair and joins in being chased. For the best part of ten minutes I can watch both raptors ducking and weaving to avoid the mobbing of up to five increasingly flustered Rooks, and making no attempt to simply leave this airspace. Looking on, I have little difficulty believing that the raptors are 'enjoying' this bout of recreation, and not just holding their ground.

FEBRUARY

The Wildlife Minister today accepted a petition signed by more than 200,000 people demanding an end to the killing of birds of prey. The petition is the largest ever collected by the RSPB. People who kill birds of prey, or consider it an unofficial part of their job description, are unlikely to desist, even if they know this. But no harm in reminding politicians of the public mood.

There aren't too many places where the public can sit back and watch Goshawks in Britain, but there is one. Visitors to the New Forest can see the birds at the nest, on a webcam in a project shared by the RSPB and the Forestry Commission. Raptor viewing is growing in popularity, and Ospreys and Peregrines will remain the A-list celebs, but I am pleasantly surprised to see that the New Forest Goshawks via a screen have inspired more signatures for the petition than any site outside London.

The top ten projects:

- Tate Modern Peregrines
- New Forest Goshawks
- Lake District Ospreys
- Chichester Peregrines
- Malham Cove Peregrines
- Glaslyn Ospreys
- Aberfoyle Ospreys
- Manchester Peregrines
- Carsington Peak District birds
- Symonds Yat, Forest of Dean, Peregrines.

Sunday

I have a picture on my wall that often attracts comment from visitors. It is a black and white, long view of a coniferous forest clearing; clear-felled, open, apart from one very tall, spindly almost branchless tree. Towards the very top of this tree – and it must be 60 feet tall – a simple chair appears to be nailed, by its back, to the trunk. And sitting on the chair is a man, straight-backed. It's called 'Self-portrait', and is by the artist Marcus Coates. Marcus's work often examines our relationship with other species. I was always intrigued that he had chosen the Goshawk of all birds to adopt for his self-portrait, although I hadn't had a chance to fully explore why with him back when I met him during an arts/science project I was coordinating from the conservation side, and he was a participating artist.

By chance, a retrospective exhibition of his work has been put on at Milton Keynes Gallery. Visiting it requires being sucked into the Milton Keynes vortex. By the centrifugal force of successive roundabouts I am hurled to-wards 'Shopping', and finally come to a halt in the retailing heart. Chancing upon its sign, I head for the sanctuary of the MK Gallery, and recover in the Marcus Coates retrospective. I find his Goshawk 'Self-portrait', and I find it oddly therapeutic.

From here I return to Steve Wheeler's, to post back the books he loaned me, and to pass the cottage in daylight, for the first time. It is bitter cold, and mercifully still. I raise the telescope with brittle fingers to look at the distant needle monument. The birds are talking about spring, if not yet sing-ing about it. Fieldfares are moving in a north-easterly direction, no doubt depleted in number after a winter on the wing over our islands. Coal Tits call – that signature note of the lonely woodland – *'pi chew'*. Mistle Thrush-es murmur, loosening up like backing vocalists backstage. Long-tailed Tits work the canopy. Woodpigeons erupt from an oak in the spinney. Shotgun reports issue from somewhere to the north. Navigating by Steve's doodled map on a sticky label I find the cottage. It is barely visible from the road, even with the trees leafless. In summer it must be well hidden, and I'm pleased that it is a comfortable distance from the public path. In the end, it was White's home for just three years, but perhaps his most significant base, as a writer.

What chance a Goshawk here, now, I wonder? How close are they? The nearest stronghold I'm certain of is probably the Forest of Dean, way out to the west. How far away is that in terms of re-colonisation – even without guns trained on them …

A check in the library of the latest bird report for the county says "two descriptions were accepted" – and not of breeding birds – for the whole year.

MARCH ~ *Sunday*

I have looked up the official records for Goshawk in my own county. It is listed under 'Uncommon species or escapes'. At the time of the compila-tion of bird records known as the first *Breeding Atlas* from the late 1960s/

early 1970s, there were precisely zero breeding records for Goshawk hereabouts. The second *Atlas*, 20 years later, had eight possible breeding records, one probable, and one confirmed. The new *Atlas*, currently in preparation, records no breeding Goshawks. They appear to have vanished again, at least as a nesting species, although the isolated sightings continue to be sporadically reported, if not accepted.

All things considered I am reasonably sure that among these local, 'unofficial' (that is non-ratified) reports are genuine and authentic Goshawks. But the element of doubt remains. How well do any of us actually know how to see and identify a Goshawk? You can read all the books in the library, absorb all the theory, but most of us have so little direct experience of clapping eyes on a Goshawk that we cannot completely shake off the element of doubt. We're just not Goshawk literate, as a rule, although experience abroad helps. Especially if you have lived there or – better still – come from there. I have a few friends and colleagues who fit this description.

Not long ago, Lars arrived from Germany to work in the International team, they of the window list and the rapid-response-to-raptor-sighting unit not far from my own desk in the Avocet building. Lars came to work on conservation programmes with our partners in Poland and Belarus. One of the projects in Poland centres on the Bialowieza Forest, probably the best and biggest remaining remnant of the primeval forest that once covered much of central Europe, and the UK. Apart from being the refuge for Bison, Wolves, Elk, Black Woodpeckers and other forest specialists now vanished from most of the continent, it is a Goshawk haven, as you might expect.

Soon after arriving, Lars went out for a walk, to explore the local area had to offer. He grabbed his binoculars, and headed up the River Ivel, the one that intersects with the River Great Ouse at my village, a few miles north from his house. He made his way towards some gravel pits, managed by fishing clubs and duck shooting syndicates. Lars was now about half way to my own home village. It's a regular haunt of mine, this stretch of the confluent rivers. He made a note of the birds that he saw as he went, and he sent them to the bird news notice-board that is published weekly at the RSPB, so people know what's been seen.

"Some bird observations from lakes east of the River Ivel, Saturday, 31 January:

3 Smew (1 male, 2 females)

3 Goldeneye (2 males, 1 female)

1 Goshawk"

A few eyebrows – including, needless to say, my own – were raised when we saw this. Who is this Lars who thinks he's seen a Goshawk? Sounds German. Doesn't he realise that we don't have Goshawks at places like this, as a general rule? Or do they know things about Goshawks in Germany that we don't know here?

We might have left it at that, and got on with our lives, and our Buzzard and Sparrowhawk and occasional Peregrine sightings. But there was a further twist. Someone else (called Tim) had also seen what he thought was a Goshawk. There came a follow-up report:

"My friend and I had a Goshawk in my garden on Sunday 1 February. Your message reports that Lars reported Goshawk near here. It would be useful to find out whether his Goshawk had jesses."

Jesses are the leather straps that falconers attach to the legs of their birds of prey, to help to secure them. Lars hadn't noticed any jesses on the bird he saw. But he might have missed them: "I didn't really pay attention whether it had jesses or not (because this is not the kind of bird where we suspect it might be an escape over in Germany, where I am from). But at least jesses were not obvious. The bird was flying about 150 to 200 metres away and was attacked by two Carrion Crows."

Tim had not seen jesses either. The Goshawk – a juvenile female – had flown through his garden mid-morning and landed in a shrub, where it rested for 30 seconds or so, partially obscured there by branches.

"Yes," he tells me now. "The Goshawk can go down as confirmed, although it was during the very beginning of my time in the UK, when I was not aware of how desperately rare they are in the UK. I think the other person said it was a young female, while I have seen it only from far. Where I lived before, as it happens in Berlin, Goshawks were quite common, and in the city I saw them every week and more often than Sparrowhawks. Well, probably because there seems to have been a pair breeding in the cemetery behind our flat."

As records of Goshawks go, they don't come much more authenticated.

"Last year," he adds, "I have also seen one while cycling in the Chilterns." "OK Lars," I reply. "Don't rub it in."

I recall another colleague who had been hosting visitors from Eastern Europe. He had been walking in a nearby town, which is just over the county boundary into Cambridgeshire. In a park, they found the partially devoured remains of a Collared Dove.

"A Goshawk has done this," the visitors calmly remarked to their host. "You can tell by the breast bone, which has these deep incisions in it". This is known as the calling card of the Goshawk. Its beak is powerful enough to tear chunks of sternum off as it eats the flight muscles of its prey's breast.

They were politely informed that it was highly unlikely to have been a Goshawk. "We don't really have them here," it was explained. They were mistaking what happens in Eastern Europe with what is possible here, it seemed. And then the bird materialised in front of them. They all saw it. Unmistakeably a Goshawk.

It occurs to me that if the visitors hadn't alerted their hosts to the possibility of a Goshawk, then it would probably have been missed altogether. The dove carcase and the hawk would barely have got a second glance. Which all makes me wonder to what extent maybe we have become blind to Goshawks, to the possibility of them. Or to have lost faith in our instincts when we think perhaps just maybe we might have seen one. Most of all I think on the whole we have stopped looking. We no longer see them, because we don't know what to look for. We are no longer Gos literate. They have been absent from our visual vocabulary, our mental reference library.

My re-fired interest in Goshawks has called these incidents once again to mind. I had kept a note of them, which is why I can reproduce the detail here. I find out a bit more from Lars. Goshawks it seems are common not only in the wider countryside in Germany, they are also resident in cities like Berlin, Cologne and Hamburg. The idea that the mad-eyed bird in the glass case, driven to extinction in the British Isles and now back but breeding only – it seems – in remote forests, could be thriving in city centres ... I have to find out more.

Lars has given me contact details for a Goshawk researcher called Rainer Altenkamp, of the University of Berlin. No point hanging around. I email Rainer to ask him about these inner-city Goshawks. He responds immediately, and couldn't be more helpful or open to the suggestion of me going there to meet him, to find out more, to see for myself. He does give me a warning, however: "This work can be cold and boring."

Rainer also mentions an expression they have in Germany; an expression that has stayed with me, toyed with me from that moment. In T. H. White's memorable phrase, it "struck fire from my mind":

"You know the Goshawk is there, because you do not see it ..."

Chapter 4

BERLIN

FEBRUARY ~ *Sunday*

I've found an interesting passage or two in a book called *Birds in England – An Account of the State of our Bird-life and a Criticism of Bird Protection*, by Max Nicholson, first published in 1926. In later life he became President of the RSPB, and is remembered as one of the driving forces of the environmental movement.

"It is not implied that there was ever an age, short of the Garden of Eden, when birds were devoid of the fear of man and would not attempt to retreat on his approach," he writes. "Like the Noble Savage period of human history, this state of things has never existed; like the kingdom of Utopia, it never will. But, although shyness is in some birds temperamental, in others it is demonstrably the effect of human persecution, and when that persecution ceases the lost confidence is almost automatically restored. Few birds retire on our approach without good reason.

"Of this I have had an interesting and convincing practical demonstration. When the Allies began the occupation of the Rhineland in 1918, all weapons in the hands of the civil population were confiscated outright, only a few light firearms for the shooting of game or the reduction of sparrows being excepted. This Waffenverbot had results of great interest and importance for bird protection."

The response of members of the crow family was the first and most obvious change noted by Nicholson. Rooks, Jays and Magpies became tame and widespread.

"Whether it was the corresponding increase of other birds and beasts of prey, or what other natural check served to keep them within bounds, I do not know, but not one of the crow family was nearly so common as in certain districts in the south of England, where the massacre of the hawks gives them a wider field and more opportunity for expansion. For game-preserving may in time eradicate the hawks, but it can never eradicate the crow and the jay. They flourish by the shooting of their less cautious predatory competitors."

And so what happened to the birds of prey in this newly gun-free, occupied Germany? According to Nicholson, "It was, in fact, astonishing. Buzzards especially became both plentiful and fearless ... in the end they began to be seen on the wing over the heart of Cologne itself ..."

And what of the Goshawk?

"The goshawk, long exterminated as a breeding species, promptly began to nest again."

***

I wake early from a vivid dream in which a hardback book has fallen from the shelf in a library. It lands open, face down, and arched. As I reach towards it, the cover of the book turns into the wings of a hawk; open, forming a kind of tent, what is known as 'mantling', when the bird comes to earth with prey and uses its wings like a cloak to conceal what it has caught, and perhaps partly for balance.

I lift the blind to squint out at the world. The blazing streetlight foregrounds a foggy, near-freezing, pre-dawn Sunday. You have to be up early to try to catch a raptor. Somewhere at the end of this fumbling around in the chilling dark might be a hawk, in Berlin. I am still adjusting to the improbability of all this as I wipe the car's fogged-up windows.

The plane takes off as the darkness has begun to thin. Autumn-sown cereals stain green the muddy fields below, beyond the gloom a featureless, flat, rain-soaked landscape. Hedgerows are reduced to thin pencil lines doodled on endless bare earth or wispy green sprouting cereal fields, trees pale shadows of their summer selves. The plane and my view of wintry lowland England are swallowed up by cloud.

Berlin is about 600 miles from my village, as the crow or passage Goshawk might fly. I wonder if German-born hawks might even make this journey from time to time, as Tim White's Goshawk did in a hamper, taken from a nest. But while Scandinavian birds can and do drift south and west over long distances, there are no records of German or other central European Goshawks drifting from there to here, between autumn and spring, following the winter migrant flocks.

For now I just want to try to find an urban Goshawk, never allowing myself to believe that I actually will. A fleeting glimpse, if I am lucky. But it will be enough just to hear more of this improbable story, from an expert, someone who knows the bird well, who is able to study the Goshawk within a short distance of his city-based university.

***

The fog still clings, adding to the impression of East Berlin as colourless, flat and resolutely urban. Spring is less advanced here. There is an absence of the birdsong that has begun to add some enrichment to our lives – Germany's thrushes are still mostly absent, on winter leave. The main signs of non-human life I note from the train are Hooded Crows and Magpies, and the inevitable Feral Pigeons.

Late morning. The sun has begun to burn through. There is a sparkle on the river, a glint on the trees, and a steady, early spring migration going on, in response to the warming air and the south-westerly winds.

A Buzzard circles daringly low over the Reichstag, the parliament building with its huge dome, shaped like an egg, redesigned in the late 1990s by Sir Norman Foster. You don't see that at Buckingham Palace, or over the duck pond at St James's Park. Jays, Magpies, Rooks and Hooded Crows are parading everywhere. Songbirds are now clearing their throats, stirring their lungs, and people are gathering for Sunday strolls. I meet Rainer at mid-day. He is a big guy, wedged into too small a car. He is conversing loudly on a hands-free as I climb in and we shake hands. "Altenkamp!" he barks, as he takes another call.

He switches the phone off, and explains that we will be visiting several different sites that he is monitoring within his research area. He covers most of the eastern half of the city, and parts of the eastern outskirts.

We arrive first at a cemetery that can best be described as poky. I can hardly believe how small.

"This is a *Goshawk* site?"

"Oh yes," he smiles.

The whole plot is little bigger than half a football pitch, off a very congested city centre street. We park and enter through a narrow gate. The graves are colourfully decorated. Some are lined with fresh sprays of pine branches, green and shiny. Goshawks do this at their nests too, adding fresh greenery through the season, which may have disinfectant value.

"I am following up on a report that this site had been recently adopted by a Goshawk," Rainer explains, all the while looking around, making full use of the available time. "Perhaps a juvenile bird in its first year of life or so, trying to set up a new breeding territory, and using it for now as a roosting and feeding site.

"Goshawks here like to bring what they have caught to quiet places, often in a tree like this' – he takes a fir tree branch in his hand – 'where they can pluck their prey and eat in peace."

Goshawks are routinely pestered and mobbed by other birds, especially when they have just caught something. Crows, in particular, will try to gang up on a Gos and drive it away, or force it off. An inexperienced or hunger-weakened Goshawk can even be vulnerable in the face of these mob onslaughts although Rainer has never heard of a physical attack. The crows are much too cautious for that.

Sure enough, we find plenty of feathers scattered below the drooping branches of a tall conifer; pigeon, crow and gull feathers mainly, with plenty of larger wing and tail feathers to collect. I help Rainer gather these, to record what his study Goshawks are catching. Sites like these will also sometimes provide him with feathers from the Goshawks themselves, which by DNA analysis or matching with feathers collected previously can indicate if it is a bird known to him, or a new one. From all this, a pattern emerges of how the Goshawks pair up, breed and disperse.

We chat about the Berlin public's attitude to having these formidable – some might think menacing – predatory birds in their midst. Rainer tells me he is not aware of any resentment, although Goshawks, especially young ones, can sometimes be anti-social.

"People sometimes call to report that a Goshawk has become stuck in a chicken house," he says. "The person who calls usually just wants someone to come and remove the Goshawk, and they are always very interested to know more about the bird."

It is possible that when this happens, the traumatised Goshawk can sometimes work out how not to make the same mistake again. Not always, but perhaps in a few cases. It is clear that some birds are repeat offenders.

"Some Goshawks are very hard to trap, others you can trap in the same way a second or third time. You can't really generalise," says Rainer.

***

We have a visitor. I sense the Goshawk before I see it. As we are talking, the alarm goes up from the Great Tits and finches foraging in the treetops and shrubbery. It is something like the murmur and frisson that would ripple through a crowd on the long-anticipated entrance of a film star. I catch fleeting sight of a dark, shadowy form hurtling over the canopy. We have a view of it for split seconds against the blue sky. It seems to have landed in the topmost branches.

I might have guessed, even had I been alone. Rainer of course knows for certain. We move carefully into position to look, to find an angle through the boughs, but can see nothing more. I think maybe it didn't settle, or more likely it saw us coming and it left, invisibly: large, clumsy ground-dwelling things with small eyes (us) versus much smaller, airborne, treetop perching thing with huge, penetrating eyes (it). No contest. But we have felt its presence. An impression has definitely been left.

In spite of all this evidence of Goshawk tenancy, this cemetery's trees aren't particularly suitable in terms of size and structure for a pair of Goshawks to eventually settle and nest, even if they may be adequate as a hideout for this youngster.

Maybe this is the only kind of view I am going to get. My expectations aren't much higher now than they had been before I set off from my Gos-free life back home, and if anything I am quite pleased to have seen it at all. That, in the end, is what Goshawks are to me. Like the Yeti, or the Wildcat, or tiger, or shark. You don't expect to take tea with them. A glimpse, a sense of it, is probably the most that can reasonably be hoped for.

**

Next stop is a Jewish cemetery, on an even busier road in an even more embedded, urbanised part of the inner city's eastern quarter. We respectfully remove our woolly hats as we enter through tall iron gates flanked by imposing walls.

In the tradition of Jewish graveyard management, or lack of it, trees have been left entirely unconstrained. They swell and tower around the headstones. The result is dense woodland, straining towards climax forest, with all the tangled undergrowth, thick foliage and fallen timber that entails. The trees are not yet in full leaf and the graveyard-cum-wood is bright and open, with shadows and dappled shade criss-crossing the tightly grouped graves and mausoleums. The whole place is carpeted and cloaked in thick, serpentine strands of Ivy. Beyond the perimeter walls apartment blocks rise above the canopy.

Rainer points out three huge nests in the highest forks of the trees. "That one is an old Goshawk nest," he says, focusing his binoculars. "And it looks now as if it is being used by a pair of Buzzards. Yes, there is one of the birds standing on it!" I can make out the form of a Buzzard on the bulky platform. It is looking back at us.

"This other one I think is a new Goshawk nest, built this year," he adds, shifting his gaze to another nest platform, closer to the apartment blocks. We both home in on it. No sign of a Goshawk.

As well as a general feeling of awe, I am struck by two things in particular. One is that the nests are so close to the apartments. This newly-built Goshawk nest is about 30 metres from a stack of verandas. It seems extraordinary, even allowing that the coming of spring and its foliage will obscure the nest.

The other thing is the proximity of the Buzzard nest to the Goshawk nest. The relationship between the two species is fascinating. Buzzards – like pretty much anything smaller than a swan – need to be wary of Goshawks. It seems that as long as they are, and don't turn their backs for too long, they may be safe enough. The same goes for Ravens. This is fit, adult birds I am talking about. Young or infirm birds are a different proposition. They need to steer well clear.

The tranquillity is pierced. Goshawk calls. Or are they? Closer investigation reveals that the sounds are coming from an unlikely source – a Jay. They can be expert mimics. This Jay is clearly very nervous, maybe of something it has detected but that we cannot, and do not, see. Maybe it is spooking itself, with all these Goshawk calls. We leave, and although there is a new nest, we are unable to confirm Goshawk breeding here.

The next site is another small cemetery, less enclosed, but in an even more inner-city location. Traffic streams past on all sides, and office blocks thrust for the sky. It has a scatter of small conifers and deciduous trees, and one conspicuously large Douglas Fir, which contains a large nest, about a metre high and another metre wide. Berlin's Goshawks often build a new nest, in the early part of the year. The male does the legwork of nest building, and he can assemble one of these edifices remarkably quickly, working most in the early morning. He gathers much of the material from the branches and canopy around him, breaking bits off with his bill. It seems even nest building is done by stealth, where Goshawks are concerned.

A raptor calls from some distance away. This is no Jay, this is the real deal. We trace the bird and, in a further challenge to my mind's eye image of Goshawks, it is sitting on the very top of a 20-storey tower block, with a grandstand view of the city and the cemetery. I am able to take a long range photograph of the bird, but it is still too far away for even Rainer to be sure if it is a male or a female, even though the difference in size of the sexes can be substantial. What is clear is the pale breast of the mature adult bird, the darker cap, and the pale eyebrow line, giving it the distinguished, prominent brow that is characteristic of adult Goshawks.

It may be calling to attract a new, or even an existing mate. I hadn't quite pictured a Goshawk sitting on a tower block, among television aerials and satellite dishes, disdaining the dive-bombing crows haranguing it.

"Goshawks live with this harassment, and mainly ignore it," Rainer assures me. "But if the mob's number gets to about 20, and they start to pull the Goshawk's tail, at that point it might decide to move on."

We talk about crows of all kinds in Britain, which exist and thrive and proliferate in our largely Goshawk-free environment, rarely having to look over their shoulder for anything more than a shotgun. There is clearly a different dynamic here, the sense of a constant ongoing negotiation between the lone raptor, like the big cat of the savannah, and the social grazers and scavengers that form part of its wide diet.

We bide our time, and explore thoroughly for other signs, but in the end we can't confirm much here either, in terms of actual breeding.

The day is drawing on by the time we get to our fourth and last site of the afternoon, in the Lichtenberg district. The bright early afternoon has lapsed into the greying murk of dusk. This site is not a cemetery, but an even more unlikely looking play-park – small and dotted with people, the general silence punctured by the strident shrieks and yelps of children and small dogs. There is a tennis court, a merry-go-round, a chute and swings. A tiny duck pond holds a few loafing Mallards. Ice is forming around its muddy rim. None of this looks at all promising. I quietly accept that I've seen all the Goshawks I'm going to see on this visit.

Rainer points out last year's nest, a thick black lump high in the crown of an enormous Horse Chestnut, like a nucleus within a cell. A little further into the park, we spot what could be this year's: another huge, dark cone in the highest fork of a Sycamore. At the same time I notice dark feathers on the ground. They catch my eye because some of them are still stirring in the light breeze, not wet and stuck to the grass like in the earlier study sites. No. These are fresh. There is a trail of them. And I notice downy feathers too, and some of these are in fact still airborne; I follow these round with my finger, I realise now I probably resemble someone in panto, gormless, until without realising it I am looking up open-mouthed and pointing at the source of this feather trail: a Hooded Crow, prone on the branch of an oak, in the firm grip of a lean, elegantly streaked, fire-eyed juvenile female Goshawk, exactly like the one in the glass case in Lamberton, only of course animate, alive in the eyes, moving, plucking and feeding, twitching as she dips her head. Purposeful, focused, alert and aware, yet somehow not looking at us. Looking beyond us. It is as though it is we who are invisible now, we the ghosts. I am breathless – never mind speechless – with awe.

"Don't point at her!" Rainer hisses. Of course I immediately feel like the slightly gauche, rookie cop, who is about to give the game away in his enthusiasm after a prolonged investigation that has led finally to the clinching encounter. I pull my hand away abruptly, sheepish.

"We need to not look at her – she might not like it," he whispers. "We should take turns to look over subtly, while talking to each other."

I can detect that Rainer, even after 15 years of this kind of work, is nearly as excited as I am. Not old and cynical like the veteran cop of cliché. Perhaps he's not encountered a Goshawk in exactly this way before. He doesn't often get this close. In a way I'm also gratified that I can still have feelings like this myself. I really do feel like the kid that once was me encountering my first close-up, wild bird of prey, a Buzzard in Mull.

And so here we are, having a rather embarrassed and stilted semi conversation. I am trying to disguise my excitement, while stealing glances at the bird. The Goshawk – the phantom of our forests, the bird you know is there, because you do not see it – plain as day, relaxed as a pet, more beautiful than books, pictures, films and of course taxidermy can ever hope to emulate – is right there before us: in a city centre playpark.

And it steadily becomes clear that she has not batted a mad raptor eyelid. And this is confirmed when a couple pushing a pram stop immediately below the branch and, as one, look up at her and, yes, point. Perhaps they too have noticed the crow's stomach on the path in front of them, discarded by the dining hawk with the bulging crop. Or maybe they just couldn't miss her. This imperious bird is over two feet long and dropping crow feathers like a gothic snow scene on a public path, after all. And there isn't that much else to look at here, if truth be told.

Freed from our neurosis about the hawk's supposed paranoia, we take photographs, and she continues to gorge herself on crow.

The peaceful scene is eventually interrupted when her mate screeches into the park. His calls to her somehow sound like brakes as he perches in trees that surround the open play area. He may be calling for a share of supper, or to check whether she has sufficient. She, after all, is the one with the far greater body mass and the role of producing a clutch of eggs in just a few weeks' time. We watch him as he races through the trees and as – remarkably

– he snatches something from a smaller nest in another of the tree crowns. I wonder if he's taken a dove off its nest. In fact, as Rainer calmly explains, he has taken an earlier prey item that has been unfinished and stashed on an old pigeon nest for later – later being now.

He carries it to the top of a domed lamp post, allowing me to follow and take photographs from a respectful distance.

We continue to watch the pair of them until she finishes eating, and goes for a wing stretch around the park and among the tree trunks. I am struck by her immense size in this context, and the stiffness of her wing beats – almost owl-like. I wasn't expecting that, from what the books have told me. But the thing I am reminded of most by her movements is a huge insect, odd as that may sound. She is like a praying mantis, somehow: weighty, heavy-bodied, crop bulging, neck outstretched, laboured wings. You can almost see the beats, as though in slowed film. It is that slow, mannered. I realise that her flight is deliberately laboured and exaggerated, for effect. She has eaten a lot, and is gestating young. But mostly I think she is showing off her prowess, her stature, her physique, the way footballers do when they are warming up – semi-slow motion movements to emphasise muscularity. It's for the benefit of her mate, mainly, but also for any other Goshawks that happen to be passing, and perhaps any other large birds with pretensions to share this small woodland cum play park. She is saying there isn't room, and that's why there aren't any crow or Magpie nests anywhere in sight. The Gos is boss here. *Gentilis* is back in the ascendancy. And as long as we are okay with that, that's the way it will stay.

As we finally leave the park in the gathering dusk, I am walking on nippy, increasingly misty, part smoky, subdued eastern Berlin air. I'm still not 100 per cent sure why, something to do with these birds being wild, yet close, forest-dwelling but content with a swing-park, impossible to see, but right there in front of us, unmolested. Whatever it is, it feels good in an all-too rare way.

Rainer and I have a bite to eat in a Turkish café on a busy thoroughfare in the city, continuing to talk at length about Goshawks. Research here started back in the 1930s, when a man called Schnurre carried out a study of the birds at a site about six kilometres east of the city. There were no Goshawks known in Berlin at that time. It is also not particularly clear whether the pesticide DDT was an issue here. It devastated birds of prey and other higher animals in the UK and elsewhere in the post-war period, until its eventual ban and withdrawal from use in the early 1970s.

By the mid-1970s there was one pair of Goshawks in the west of the now partitioned city (the Wall went up in the early 1960s). There may have been one or two pairs in East Berlin at the time as well, but this is not certain. A decade later there were 16 pairs in the East, and the Brandenburg district beyond. Today, there are 90 pairs, and the city is thought to be at capacity, with roughly 10 pairs for each 100 square kilometres. The factor that limits there being more is not the availability of prey species – because that remains abundant – but because the Goshawk needs a bit of space from its neighbours. They are inclined not to come to blows – that would probably mean mutually assured destruction, given the weaponry at their disposal – but they do represent a threat to and can depredate each other's young. This keeps a lid on the population. To increase in number, they have to disperse outside of the city boundary and find new or vacant territories elsewhere.

Other German cities have also now established thriving Goshawk populations, including Hamburg and Cologne. Munich, meanwhile, has none. Rural Bavaria is still less enlightened about the Goshawk. 'It takes a long time and many generations for the Goshawks to become accustomed to people', Rainer tells me. It may be that the braver birds are the most readily killed, which feeds back into the population, and makes some Goshawk populations especially reticent. Perhaps something like this is happening with the UK population, making the birds unwilling to colonise new territory types, even though these have plenty of food and suitable nesting and sheltering sites.

Goshawks have some interesting impacts on the things that they depredate. Crows and pigeons remain abundant in the city, and reproduce far faster than 90 pairs of Goshawks and their broods could ever hope to make an impression on. But the presence of Goshawks influences where other birds can choose to nest. When Goshawks nest in a park, there tend to be no Hooded Crow nests there. In a year when Goshawks chose not to nest in one park, there were 14 pairs of crow nests in that same park – a remarkable density. With Goshawks present and breeding, the crows have taken to nesting in tree-lined streets. Sparrowhawks and Magpies must also avoid the vicinity of Goshawk nests. I have also noticed a Rook roost by one of the stations downtown, in a tree hemmed in by tall office blocks and traffic noise. I wonder if this might have been in response to Goshawks being phantoms of the park, in the modern day. It could also be a micro-climate thing.

Several bird species are decreasing in the city, which might lead one to be suspicious of the influence of the Goshawk. Collared Doves have dwindled to a core of about 80 pairs, from around 10,000 in the early 1980s. Jackdaws have almost gone completely. Ring-necked Parakeets, which have exploded in number in and around London in recent years following introduction from their native Asia, have gone. This is despite unofficial attempts to introduce them. The Mistle Thrush also seems to be on the way out.

Meanwhile, other species that are routine Goshawk prey are steadily increasing. Woodpigeons, Starlings and Mandarin Ducks are all growing in population, while the Peregrine Falcon, with which Goshawks might spar and accord mutual respect, is also doing well. A pair nests on the Rathaus – or city chambers – at present, also right in the heart of the city.

In our cafeteria debriefing session we also cover some of the other wildlife here that is coming back, or has been hanging on, including the Wolf. Rainer explains how a Wolf can survive in the unlikeliest of places, as a solitary animal, keeping a low profile, scavenging and predating small prey, waiting to find other packs. Wolf packs are not static units, they must disperse and mix with new blood. The pack is just one part of the life-cycle. We speak also of the White-tailed Eagles that are nowadays regularly seen within the city limits of Berlin; huge birds, growing increasingly accustomed to people, and tolerant of our proximity, as we are to theirs.

Rainer wants to check the leg rings on the Goshawk we watched. We arrange that he will pick me up again tomorrow afternoon, and we will return to that unlikely scrap of a public park to find the Goshawk again, and – he reckons – to catch her.

Monday

There is no sunshine today to lift the morning's enduring gloom. I am glad I packed gloves and hat. Rainer is similarly wrapped up as he lifts his Goshawk-catching equipment from the boot of the car. This consists mainly of a small cage, and a cardboard box. Inside the box is a perky, ivory-white Barbary Dove, which is quite calm, and remains so throughout what follows – as peaceful as its image suggests. The dove is the lure, and will be placed in the cage. I am assured it will be quite safe. Rainer has been looking after and deploying this same bird for two years and dozens of missions. You might forgive it for being nervous, recognising what lies ahead, but then again it is accustomed to living through these experiences. It really does have no reason to fear for its own safety. And it is lavishly provisioned with seed.

The trap works by having cat gut – or fishing line – nooses tied across it. Any bird landing on the trap gets caught by the feet. Traps like these are easy to set and straightforward to release captives from unscathed, and therefore an effective method of catching a Goshawk for research purposes. Rainer has caught nearly 70 in the course of his studies. Some he catches within minutes, others can take all day, or even longer. I now understand his earlier warning to me to be aware that this work can be tedious – and cold.

With all that experience, it takes Rainer a surprisingly long time to decide where to site the cage within this play park. Although there are very few people around on a weekday, he knows from his many hours in 'the field' that he has to strike a balance between the cage being visible and accessible to the Goshawk, and for us, but not so to passing dogs, children and curious adults.

Having selected the likeliest site – away from the main footpaths, but in the middle of the open area in the centre of the play park – we retire 100 metres to await developments. We wait. And we wait. And we get cold and stiff, and aching in the lower back. I take a number of strolls to re-start my circulation. We watch dog-walkers and pram-pushers come and go. An elderly couple passes us. They pass us again, about 15 minutes later. In total they will pass us five times. It confirms just how small this park is. They do this many laps to turn the circuit into a walk of decent length. By the final time they pass us, I feel I know them, and they are smiling benignly, though they don't ask us what we are doing, oddly. Goodness knows what we look like. A couple of undercover commandos, I suspect. I'm not sure we would have been able to do this work 20 years earlier in Cold War era Berlin, at least not without a thick wad of documents in our inside pockets.

Our female Goshawk is suddenly with us, calling overhead somewhere beyond the trees. Rainer spots her as she glides through the branches and comes to rest out of sight. She soon moves to a more prominent branch, high in a bare-limbed Linden tree (we call them Limes) over to our right. From there we take it that she must have a clear view of everything in the place, including the dove, and it must be only a matter of time ... more of which passes. Doubts creep in. What if she isn't hungry – still full of crow? Or isn't in the mood. Or is suspicious of us, or others in the park. Or maybe, in spite of our painstaking selection of the site, her view is obscured.

For minutes, and then hours, she continues to sit there, no more than a distant bird shape – highly visible if you know where to look, and what for. But not doing much. Thoughts of Gos meander. I wonder, if I had happened to see this bird up there, on another day, would I have had any clue it was a Goshawk, or even looked at it twice? I might, but without binoculars I wouldn't be able to tell it from a Buzzard. And until I found out about these Berlin Goshawks, I wouldn't have had any sense at all that Goshawk was a possibility in such a place.

Rainer shuffles, impatient. He has begun to question if she can actually see the dove from up there. Perhaps Goshawks can't see through walls – or trees – after all.

"Would she not have seen it when she came in?" I ask him. "She seemed to fly right over it?" How could she not, with eyesight like hers? But maybe her focus was on other things.

While she may not have seen it, our trap hasn't gone unnoticed by other, non-target, mega-fauna. We watch anxiously as some of the park visitors venture near to have a closer look at our contraption. We don't want to have to intervene, as we don't want to risk putting the Goshawk off. But there is no choice and Rainer has to jog back over there. A young woman pushing a pram, with a young child in tow, has stopped right by the trap and is staring down at it. Rainer explains to her what it is for. Like most of the bystanders he's had this kind of conversation with, she is intrigued. Well, it's not what you expect to find in your local play park on your morning perambulation with the kids. He's never had any issues with anyone over it.

Because he's had to go over to the trap anyway, Rainer figures he might as well move it 10 feet, to where he can be sure it is in the Goshawk's line of vision. He then retreats and re-joins me, calmly walking back through the central area of the park, under the tree limb on which she was so gloriously enthroned yesterday, and across the damp grass to where I am, by the bench.

The Goshawk calls, probably to her mate, who remains unseen through all this. He may be in town, hunting, or loafing, like her. Boom, and loaf. That seems to be the lifestyle. And then she flies off. We watch her disappear like the phantom of the books, into the mesh of branches, the vapours of the inner city beyond. Our spirits sink. And then a very curious thing happens. 'She is on the trap!' Rainer exclaims, and he is already back into a half-jog, needing to get there fast, but not too fast, not before we can be sure she has been caught by the feet. I can see her broad, stick-coloured wings flapping in the arena as she half falls sideways, the trap lifting below her onto its side. Rainer is now full-on running the hundred metres. I am running too, in his wake, getting camera poised as I go.

It must only take us 25 seconds to get there, but in that time the most extraordinary thing has happened. This silent park has been transformed into a scene from a horror movie: a din of crows going berserk in the air around the flailing hawk, and the two watchers who are now gathering it up. I am sure that many times a flock of birds has been described as like a scene from Hitchcock, but I have never seen anything quite so much like it, and am bamboozled by how so many crows can have been close enough by to get here so quickly. They have materialised as though from the misty air of the park. I am still trying to fathom how the Goshawk got to the trap without us even seeing her.

We work out later that she attacked by flying at right angles to the prey, as though away from it, through the trees, to get low down, then she came at it just a few feet off the ground, presenting her thinnest profile to maximise invisibility, as well as pace – a deadly will o' the wisp – and a proper insight into just how the phantom moves.

And all the while at least one crow must have been monitoring her, perhaps from a rooftop some distance away, and given the call when she struck, partly out of genuine alarm, and partly perhaps because a dozen crows might yet have a chance of driving away a Goshawk, even a female. But not often. And not today. We are the ones doing that. Her beak is partly agape and her yellow-orange eyes are flashing when we reach her. I'd describe her state as halfway between the wild, free bird and the stuffed one in dramatic, man-made pose. And if my heart wasn't racing before, it certainly is now.

While we were waiting Rainer had been giving me a few theory lessons on how to handle a Goshawk. Now the practice kicks in. I am nervous, partly to do with not wanting to prolong the hawk's trauma, and of course desperately afraid of damaging this beautiful animal, but also because her talons – of which I am in charge (this is my main responsibility – to manage these while Rainer takes her measurements) are an inch long: the hind claw and the main foreclaw, at any rate. Literally inch-long semi-circles of jet black, tapering lethalness. A Leopard's can scarcely be any bigger.

"What about her beak?" I ask Rainer. Holding her legs is one thing, but these aren't her only weapon – clearly. 'Goshawks don't bite,' he tells me, probably not as reassuringly as he thinks. 'What do you mean, they don't bite??' I am replying, dubious. Why wouldn't they, after all, in self-defence, as they do in dismembering prey.

"What, never?"

"Well, not never ... "

I guess he has to hedge his bets a little.

Beak threat or no beak threat, we have to work quickly. Rainer issues brisk instructions, most of which I repeat after him before acting, just to make sure nothing is being lost in translation or garbled in our excitement. We are new as a double act, after all. My hands are cold and clumsy. I think the blood has further drained from them in anticipation of possible mishaps. A rare blood group, I might add. The whole operation takes about ten minutes but I am willing it to be about two, and it feels like 20.

First Rainer notes her leg ring number. (Later checking reveals where she was hatched and reared last year – we already know she is a first-year bird, from her plumage.) Rainer measures all her vital stats – wings, tail, beak, feet. He marks all her major wing and tail feathers, so that if and when any of these are recovered, from a nest or roost site, especially at the time of the moult during breeding, when birds shed their plumage and grow a fresh set, she can be identified and linked back to this site. She must be weighed, and for this bit I must release her legs, and hope to get them back again safely in my careful but firm grip when she is removed from the weighing bag. She weighs 1.33 kilograms. To give an idea of what we are dealing with here, a female Sparrowhawk, at 300 grams, is well under a quarter of this weight.

Before he releases her I am able to take a few quick photographs. She glares at me – I wouldn't even call it defiance. It is something more, or perhaps less, than that. Knowingly, perhaps. Determinedly. I can hardly see any extra emotion there from her, just her constant, indifferent, almost contemptuous haughtiness. It is as though she is not afraid. Too mad for that.

Constantly mad. And invincible.

But it is a fact, I can confirm from this one experience of handling a wild Goshawk, that she never showed any inclination to want to try to bite either of us. Nor did she at any stage moan, or struggle, or call out, apart from one strange little yelp, or exhalation, as she finally realised she was free, laid out on her front on the grassy lawn of the park. Muscular, long-bodied and stiff limbed, she spread her wings, kicked out her feet, and was gone in an instant, like a high-powered aircraft negotiating a tight space. Goshawks aren't the most difficult birds of prey to handle. Falcons are much more feisty, despite their smaller size. And they do try to bite you, as well as claw, and struggle, and scream. It is just as well the Goshawk doesn't do this. It is just a peculiarity of their character; and it is odd then to reflect that they can be straightforwardly warlike in defence of their nests. Rainer, when he climbs tall trees like these to ring young in the nest in late May, wears a full riot gear helmet, and a special outfit that enables him to withdraw his arms to his sides when the irate and understandably protective parent Goshawks come in to rake him with their scimitar claws. That is a fearsome adversary when you are strapped to the topmost branches of a tree perhaps 20 metres off the ground, with limited movement and visibility in that headgear. Rather him than me. Rainer sometimes consults a tree-climbing expert when assessing whether a nest is reachable or not. He doesn't attempt them all, by any means. He isn't sure about the nest that our Goshawk has built with her mate in this park. He will have to consult.

Tuesday

I have time alone to explore Berlin on a hired bicycle, on its network of cycle lanes. I soon find my own Goshawks, with one calling near a nest beside the Zoological Gardens. I find a pair active at a nest in a numbered tree, with another nest nearby, in part of the park near the Philharmonie. I get chatting to a few passers-by. Some are not aware of the Habicht, as they call it here, but are fascinated to learn about the birds. One lady knows them so well she can tell me that the other nest is last year's. Rainer later reveals that this was in fact the nest from which our female was reared and ringed a year ago.

I sit down for a while in the park, to watch ducks, crows and pigeons milling around. Rainer told me that if you watch a group of pigeons in a Berlin street for one hour, the likelihood is you will see a Goshawk attack. He also explained that Magpies are especially vulnerable if they leave the safety of a tree. Their long tails and short wings are designed for tight spaces. In the open, they are quite simply ill-equipped and too slow and clumsy to avoid a Goshawk that has locked-on to them. In a tree, they can often make good their escape, by playing chase around the trunk. The Goshawk will usually give up.

Towards the end of the day I find another juvenile female by a nest in a tall tree near the Bismarck monument. At dusk I watch a male Goshawk hunting in another part of the Tiergarten. As darkness gathers, there is one final, magical moment. House Sparrows are queuing to roost in the barrel of Russian tanks installed here as monuments to the end of the Second World War, after the Red Army had rolled into the city. A fitting gesture of peace in Europe now, in our time. Just across the way from here the cornered Adolf Hitler, run to ground at last, took his own life and was burned in a bunker, near the Reichstag.

**

Back home, the image of the Goshawk in the play park endures, almost effacing that of the glass-eyed bird in the cabinet. I have turned over in my mind many times how this bird can be so at home in Berlin's noisy, dusty, concrete, exhaust-filled streets. But it's clear. The bird has all that it needs there, in the city centre: tall trees to nest in, plenty of food to eat, physical terrain that suits its foraging techniques, no one harming it.

I have never had a bird of prey – or any bird, really – encounter of this vividness, this magnitude. The Red-tailed Hawk in Central Park and a

Buzzard in the Hebrides come close, but for me the Goshawk, radiating grandeur, brimming with attitude, aglow with life and purpose, is on another level. Above all, Berlin has left me blowing on the embers of the nagging question to which I need to find an answer: where is our Goshawk?

Chapter 5

ENCHANTED

MARCH

This house faces south. The bed is level with the windowsill, so I can view
the world propped on one elbow or a couple of pillows. It's a typical
slice of rural, lowland Britain – hedgerows, large arable fields, smaller fields
for livestock, paddocks for horses, farm outbuildings, hawthorn hedges,
scattered oaks, spinneys and even a woodland about half a mile away, to the
west. The Enchanted Forest, I call it, optimistically.

The houses most recently added to the northern edge of Sandy are now
visible on the southern horizon, three miles away. The streets and closes in
these new builds are all named after birds, in recognition of the town's association with the RSPB, which has its HQ on the greensand ridge that bolsters
the eastern horizon.

It is by most standards a flat landscape, but by no means featureless,
and the skyscapes compensate in part for the absence of serious hills, far
less mountains. I revere mountains, but I also love that the sun shines on us
a lot here in eastern England, and its progress across the sky is unobscured

by landforms or trees that are too close. I also crave forests: being in them, exploring them and imbibing their atmosphere, but at the same time I prefer day to day life in open country. Space and light are vital. I don't think I'm unusual in this, and the history of our relationship with woodland as a species in northern climes supports the theory.

I can watch this view for long minutes at a time, enjoy it as it changes through a day, if I'm writing here, and through the seasons. My wife Sara will catch me at this, and sometimes urge me to get outside. She tells me I remind her of her old cat, or a man she used to pass on a street, who would stare from the window, and wave, each day. He had lost the use of his legs.

I particularly like watching the progress of birds across this sky, and I am struck by just how many birds there are around here, in numbers if not variety. Crow species in particular are abundant. A bustling and expanding rookery lines the great north road, a mile to the west beyond the wood. Jackdaws join forces with the Rook flocks to form huge roosts there. Carrion Crows and Magpies are never far away, the latter often bouncing into the front garden for scraps of food, and scraps with gangs of Collared Doves, when these are trying to nest.

Some people moan about Magpies, and I think it isn't just for their distasteful habit of eating other birds' babies. No one ever held parasitism and infanticide against that national treasure, the Cuckoo, after all. I think there is something in the apparent arrogance of a Magpie that grates on people. Something about its cockiness. The glint of knowing mischief in its eye. They can look invincible. Unfettered. Too clever by half. Perhaps they largely have been, in the world we have created for them.

The odd Jay, that other dandy of the crow family, pops out of the woods to visit, but they never look truly comfortable away from the sanctuary of the trees, perhaps because they don't fly too quickly. But in the absence of Goshawks, what have they really to fear apart from Sparrowhawk ambush? Collared Doves form energetic flocks in the road, and nest on drainpipes and even on telephone wires, where these radiate from central poles above the pavement. Woodpigeons have proliferated in recent times, and seem to get tamer by the year, nesting, like the doves, through eight or more months of the annual cycle, including in the back garden, on flimsy twig saucers you can see through from below.

House Sparrows visit in conferences up to 40 strong, and roost in the tangle of Golden Hop and Wisteria that cloaks part of the front of the house. Starlings can number a hundred on a June afternoon, fragmenting thereafter and visiting in smaller bands. Gulls form great flocks in the autumn as the crop stubbles are ploughed in. Lapwings gather and moan through the night in the field behind the house. The sky above is a regular procession of visitors from north Europe: geese, Redwings, Fieldfares. Herons, swans and Cormorants sometimes beat over. Swallows and House Martins are a constant aerial presence in summer, joined by Swifts in the upper stratum later. Beyond them, the vapour trails of plane traffic into and out of London airports score the sky to the distant south.

From this distance the Enchanted Forest looks much like any of the small islands of woodland that decorate the arable landscape of central, south and eastern England. These places are often out of bounds, inaccessible, reserved for autumn and winter pigeon and game bird shooting. So we are lucky with ours. I can usually see the crows and pigeons skirling around it. Sometimes it produces or absorbs a bird of prey. It's too far away for me to make out much more for certain than that sometimes the bird in question is a circling Buzzard.

"It doesn't look very enchanted to me," said Sara, when I first brought her here. It's possible I over-sold it. To be fair, the particular part of the 'forest' we were in at the time *can* be a little dead-looking, this being early spring, and with the leaf buds not yet unfurled to conceal some of the worst of the jetsam pushed out of cars by passing motorists. Esme Wood is its more prosaic name, on the map. Local people still know it as Home Wood.

My hype is justified because just under a year earlier, among these trees, I stumbled, in the arrowing rays of a sinking sun, upon a pair of Badgers cavorting 'neath a spreading chestnut tree. At that moment, I, if not the very 'forest' itself, was definitely enchanted. I explain all this to Sara, enjoying the memory as we pass the now quiet sett entrance. She follows faithfully, if not entirely convinced.

Luckily for me and my wood credibility, further enchantment is just up ahead. We both hear the rustling of leaves on the bank of a shady pond. A furling Grass Snake. It goes into reverse and shrinks away at our approach, crinkling across dry leaf litter as it takes cover. It slides out of view under a stump, where the first new leaves of Nettle and Wild Arum afford further concealment. We sit down close by to await developments, assuming the shy serpent will be driven to re-emerge, to seek some more energising sunrays after a long winter in cold storage. That, after all, is what we are doing on this, the first properly warm day of the year, with Chiffchaffs re-invading the tree-tops in hearty voice, and buds easing out of their casings all around.

We can hear snake skin bristling ever so slightly across crisp debris, so we know it is still in there, still active, and quite likely to re-emerge if we are patient. Sure enough, a small angular head appears, and a flickering black tongue tastes the air in our direction. Then another head emerges at the other side of the Nettle patch. There are two snakes. The head of the first one stretches slowly towards us, wobbling ever so slightly on a narrower neck, hairline-thin tongue forking the air.

Snake 2 then slides boldly out from the lair to lie across the leaves and twigs, to soak up some more warmth. Over the course of the next few minutes, to our delight and amazement, a third, then a fourth, then a fifth snake make their way out of the hibernaculum towards the water, and weave in turn out through some submerged branches and into the duckweed. The largest of the snakes is perhaps a metre long, takes to the water and swishes with great elegance across the dark pool, head held proud of the surface, yellow collar prominently displayed, slicing a thin wake in the duckweed as it goes.

I've always loved the magic moments that spring renewal provides, but I've never witnessed an emergence quite like this one. Even someone with an instinctive fear of snakes would have been impressed, I am sure. Our encounters with snakes nowadays, in these islands, seem to me so infrequent it would be hard to experience one without the sense of awe and slight incongruity that we experienced on this faintly surreal occasion. This innocuous little stump bore forth a Medusa's head of serpents in an otherwise quiet (bar the warblers above and a complaining Moorhen) and familiar place. I know this wood well, but I've never seen a snake here before, let alone five. Is there a collective noun for snakes? A hydra, perhaps.

Snakes are enchanting. I love the economy of their design, and the ease of their movements. They make other creatures look fussy, complicated, overproduced. But they also seem somehow naked – exposed – especially in a yet-bare woodland.

So, even before the possibility of Goshawks here had occurred to me I had begun looking more closely at the wood, and the spinneys. Even these have thrown up surprises. One conceals a large, iris- and rush-lined pond, with a Moorhen nest in a straggling willow, and a pair of Mandarin Ducks, little Chinese dragons, in a muddy ditch. This, I worked out from an 18th century map, is all that remains of the 'Great Marsh'. The 'Mashes', Mr Bettles still calls it. The Great Marsh may be reduced now to a little sedge-fringed pond within the field network, at the intersection of a grid of draining ditches. It's gratifying to discover that it can still throw up a few surprises.

Wednesday

Some people have queried whether the Goshawk is native to the British Isles. A recent book – *The History of British Birds* – helps to nail this. Authors Derek Yalden (University of Manchester) and Umberto Albarella (University of Sheffield) estimate that there were 14,000 pairs of Goshawks here in the post-glacial period. This is very important, because some authorities – and some people antipathetic to the Goshawk's interests and status here – would like to believe there were none at all.

Archaeological digs often turn up bird bones, with all the other artefacts. For many years records of these have lurked in archives, labelled generically, gathering dust. A few years ago, Yalden and Albarella set up a research student to find, sort through, analyse and log these records. I have been thumbing through the index of the resulting book to pick out the Goshawk entries. There are many.

The book also throws up some intriguing questions. Take Orkney. It has long been known and it is often cited that the White-tailed or Sea Eagle must have been of great importance to the early people there, because its bones have been found in the so-called Tomb of the Eagles, which is legendary. It seems the birds had enormous totemic significance to have been given such a place of honour.

I am intrigued to learn that Goshawk and Tawny Owl bones have also been found on the islands. The book reveals that "somewhat surprisingly, given the treeless nature of Orkney, Goshawks at both Howe and Skaill … (Iron Age) … a Tawny Owl, perhaps a wind-blown stray, seems as unlikely on Orkney as the Goshawk". It's possible both species remained resident, but Passage Goshawks we know pass through and over Orkney, even today, on occasion. These are most likely birds from north Europe. Mesolithic people may have devised some method of catching them, unless they brought them in by boat. But for Tawny Owl it's a different matter. They don't fly anywhere that's across sea, as a pretty hard and fast rule. There has never been a record of the species in the islands, although they are present just across the Pentland Firth. They simply cannot cross the water between mainland Scotland and there. So how did the Tawny get to Orkney and, more to the point, why? I wonder if it could be linked to the Gos records.

I'm not an archaeologist, but it makes me wonder if these bone records might indicate a captive owl being used to lure passage hawks in to traps, as has often been practised at other times and places.

I've been in touch with the authors. "The logistics of humans getting a Tawny Owl to Orkney would have been trivial," Derek has told me. 'They got sheep and cattle there about 5,500 years ago. They also got them to the far west of Ireland about the same time. They were better sailors, and had better boats, than we know.

"The fact that both Tawny Owl and Goshawk co-occur at Howe is not especially significant, given the long bird list from this site, though it might be relevant," he adds.

Orkney had low tree cover of species like birch, Hazel, alder, willow around 8,000 to 5,000 years ago. A combination of climate factors and Neolithic farming saw most of it disappear since then.

I get a view from Umberto. "Your idea is entirely plausible," he tells me. "Tawny Owls may have been introduced by people. As you say, the use of Tawny Owls as decoys is well known, though unfortunately I cannot think of any historical record that mentions this in relation to Britain – but this doesn't mean that it doesn't exist."

Intriguingly, Goshawk bones have been found at the Tomb of Eagles itself. I wonder how those got there, and why. Might the birds have been put to good use, while alive? We know that falconry was learned in later times, from the east. The Saxons are the first to have definitely practised it in the British Isles. But I'm wondering what would have stopped people in Orkney at that much earlier period pretty quickly working out the value of a Goshawk as a seabird catcher, especially outside the seabird breeding season. Anyway, always fun to speculate. The answers, if they are to be found, might hang on interpretation of the fine detail of the artefacts and remains found during the excavation.

The Domesday Book, the audit of everything in England and Wales carried out for William the Conqueror, and completed in 1086, contains reference to Goshawks, and indicates just how highly valued the birds were at that time. The book includes a count of 24 Goshawk eyries in Cheshire, including Macclesfield Forest, all appended to woodland entries, and with a value of £10 – a princely sum then.

"Both the value and the listing are a clear indication of just how valuable

Goshawks were to Norman aristocracy," Derek has told me.

Many old references to the Goshawk are unreliable because of possible confusion with Peregrine Falcons. Not only were Peregrines sometimes known as Goshawks too, but of course not everyone could be trusted to know the difference between these species (or to consider the difference important). Derek is comfortable these Domesday Goshawks were the real thing. "These hawks clearly were not falcons, which would not be breeding in these forests."

He also points out another challenge to the received wisdom about Goshawks, which is that in the hierarchy of things, they were the birds that yeomen were permitted to own, in the strict caste system of bird of prey ownership set out in the *Boke of St Albans* from around 1480. The *Boke* is also known as the *The Book of Hawking, Hunting and Blasing of Arms*, to give an idea of its content.

"Goshawks were also much too expensive for yeomen to own, contrary to received wisdom from information contained in the widely cited *Boke of St Albans* (which gave us a Kestrel for a knave, etc),' he says. 'Yapp (1982) pointed this out, and is obviously right."

***

It is commonly believed that the Goshawk was first reintroduced to some of the remoter forests of Scotland from the late 1960s, but in fact there were Goshawks at large in woodland much further south, no distance at all from central London, in fact, from a much earlier date. The presence of this pair, of which at least one was without doubt an escaped bird, was a closely guarded secret.

Author and naturalist Dick Orton describes them in his book *The Hawk-watcher*. He learned of the birds through contacts in the army, and of the attempts made by Lord Alanbrooke to photograph the birds at their nest in the early 1950s. I have discovered that the famous bird photographer Eric Hosking had also been in on that mission. In his autobiography, *An Eye for a Bird*

(Hosking lost an eye to a Tawny Owl), I find the following account of this:

"One day in April 1951 Lord Alanbrooke telephoned to say that Lord Portal (Chief of the Air Staff during the war), an ardent falconer, had invited him to go to Cocking, near Midhurst in Sussex, to see the nest of a pair of goshawks and he wondered if I would like to accompany them. I did not even know that there were any goshawks nesting anywhere in Great Britain, so the news excited me and I enthusiastically accepted the invitation.

"As we walked through the wood I wondered whether the nest would be in a position where we could take photographs. Most of the trees were enormous beeches, some standing more than eighty feet high, and almost at the top of one of these was the nest.

"A forester climbed the tree to find what the nest contained, and as he swarmed up I kept a careful watch on the nest as I particularly wanted to see the bird leave. The climber had almost reached the nest before the bird suddenly leapt off, and as she flew through the young, fresh, green leaves I could clearly see that she trailed jesses."

Disappointed that the bird had this appendage, and perhaps mainly for that reason, Hosking's enthusiasm for the project waned.

"Most of the thrill of photography was lost for me by knowing they were escapees, but apart from this the nest was in such a difficult position I do not think it would have been possible to erect a hide. The nest contained three eggs all of which hatched and the young were reared. I still have a feather the forester brought down for me."

He went on to photograph Ringed Plovers instead, so these fugitive Goshawks in our southern midst weren't captured on film. But what became of these early Sussex Goshawks? Robert Kenward thinks they were discontinued soon after Hosking's visit. He cites Richard Meinertzhagen writing in 1950 and 1959 as his source for this insight: "colonisation by two breeding pairs in Sussex in the late 1940s failed, with the disappearance of the hawks shortly after it was reported that they were eating mainly pigeons and pheasants."

Tuesday

I would love in this investigation to be able to describe finding Goshawks breeding near my home. Even better, to be able to watch them, and describe their comings and goings. Realistically, I am unlikely to find such a thing. And even if I do, there is the small matter of the near impossibility of discovering Goshawks without them seeing me first, and being disturbed. Apart from anything you need a licence for such endeavours.

Paul Marten is someone for whom all of the above falls into place. He has found a Goshawk nest in his home county, in such a place that he can watch it without inconveniencing the birds. He has the necessary time on his hands to get up at or before dawn each day to install himself even before the birds are conscious and active, although he can actually access his viewing 'hide' without them seeing him. He has been able to spend many hours waiting and observing them, and – crucially – he is licensed to do it.

The other good news is that these Goshawks are – like Eric Hosking's – in Sussex, southern England. They are back, but of course the exact location has to remain secret. Paul has kept a detailed website diary of his experience, that I have quoted from below, with his blessing. Although he believes Goshawks have been present in the area for some years now, he thinks this is the first proven modern-era breeding of the species in Sussex. The nest is within driving distance of his home and he has been able to spend many hours observing it between late March, when the nest was built in a larch, and July, when four young fledged. A conveniently sited clump of smaller spruce trees has allowed him to watch the nest unseen by the birds.

'When I crawled underneath them, it opened up into a sort of cave. From here I could sit quietly and very comfortably, and watch the nest through a 6-inch opening I snipped out of the foliage, from a distance of about 80 feet, and the birds never knew I was there. As a testament to how good my cover was, I never heard either of the adult birds give an alarm call or show any sign of agitation or nervousness in the whole time I was present. Other bonuses to sitting silently under a Christmas tree in full camo, for hours on end, were the pair of Firecrests that would hunt by my head, a male Muntjac Deer that walked within six feet of me one morning, and a Fox that actually came up and sniffed my boot early one day. These interesting interludes were very welcome as most of the nest observation during the incubation period was just that ... staring at a pile of sticks! Once every four or five hours the female would shuffle around a bit and, if I was lucky, she'd stand up briefly to turn the eggs. But for the vast majority of the time all I would see was her tail sticking up from the top of the nest."

Paul describes the narrow escape of an unsuspecting Grey Squirrel. "He came down the trunk and walked onto the nest where the female was incubating. I held my breath ... he stopped dead in his tracks, did a really funny, cartoon-like double-take, and his eyes almost bulged out of his head as he realised where he was, and what was sitting in front of him. Before the Gos could react, the Squirrel just leapt straight out of the tree and fell 45 feet to the ground and sprinted off through the wood, faster than any squirrel I've ever seen."

Paul came to know the female as Heidi, and the male as Casper, for his ghost-like appearances and disappearances. Heidi seems to have been towards the more aggressive end of the Goshawk personality spectrum. 'When Casper wasn't hunting he was usually sat somewhere in the wood near the nest, but he *never ever* approached it, while Heidi was at home ... He would pass over the nest like a bomber, and drop the prey onto the waiting young. Even this innocuous action would infuriate the female, who would chase him out of the wood, screaming like a banshee. She simply would not tolerate him near the nest.

"I would sometimes see his shadow above me as he glided silently onto a perch, or he would call to Heidi to let her know he was around, his piercing '*kek...kek...kek*' raising the hairs on my arms whenever I heard it. I had one magical day when he actually landed in a tree where I could see him through my scope, and I watched him preen for 37 minutes ... wonderful!!!

"Casper took all of his kills to an oak tree at the edge of the wood, and plucked them there. He would take the prey, usually a Woodpigeon early in the season, eat the head, pluck the rest of it and then call to Heidi with a soft, very quiet '*guk*' note, and then fly off. Heidi would then slip silently off the nest, fetch the prey from where he'd left it, and bring it back to the nest and eat it. She was never away from the nest for more than a minute or two. It was normally after she'd eaten that all of my sitting silently watching sticks was rewarded, as it was at these times that she would sit on the side of the nest and enjoy the sunshine whilst surveying her realm. She once sat like this for three hours. Bum ache and cramp are soon forgotten when a wild Goshawk is sat in front of you."

As the chicks grew, the female was able to sit in neighbouring trees, supervising from a distance. Paul feared for the runt of the brood, but it put on a growth spurt and caught up with its siblings. Gos chicks are surprisingly gentle with each other compared to many other raptor and owl species. With Heidi's back no longer turned, Paul would definitely have to take up position within the spruce cave before daylight.

"There had been talk earlier on about ringing the chicks, and I was asked if it could be done, but I said no. The birds are beautiful enough as they are ...

"Once the young had all left the nest, most of their time was spent sitting in their father's plucking oak, waiting for either Heidi or Casper to turn up with food. When they did, it was absolute bedlam, with all four young fighting to get the prey item from the parent bird. Five Goshawks fighting one another really is something to behold."

With the young safely raised and dispersed, to make their way in the world, Paul was able to examine the leftovers. Among the huge number of pigeon, Magpie, Jay and other crow bones and feathers below the nest and the plucking tree, he counted 22 Grey Squirrel tails, and even Sparrowhawk feathers. Interestingly, there was not a trace of Pheasant. "Despite there being a lot of pheasants being reared very close by, I didn't find a single bit of evidence relating to pheasants being preyed upon. In fact it would appear from my findings that 90 per cent of this pair's diet was made up of all the species that gamekeepers spend a lot of time trying to keep down."

April

There have been developments locally. My colleague Peter Newbery, a man who ought to know, reports seeing a Goshawk in the town just five miles up the road from my village. "I know this sounds a bit stringy," he writes (birder speak for over-claiming on something possibly seen) "but it had all the hallmarks of a Goshawk, and was chasing pigeons over the market square."

I contact Peter straight away. "I'd be interested to know a bit more detail – exactly where, etc. I've been following up Gos reports as part of an ongoing investigation!"

"Not much more to say, I'm afraid," he replies. "I was walking through the Market Square when a large raptor flew over the rooftops and stooped at a small party of Feral Pigeons further down the High Street. A brief view but it looked just right for Gos. Hope this helps."

By coincidence, on Saturday, a day after Peter's sighting, I had been walking on some local Heathland, and had noticed, a long distance away, a raptor stooping at what I assume were Rabbits, on a steep slope crowned with pines. My impression was of a light-brown bird, no more than a streak, then gone. I had put it out of my mind, but now I'm wondering what it is that I might have seen.

Thursday

The Guardian has published a short piece I've sent them on Berlin's Goshawks. As though to confirm our general unfamiliarity with *Accipiter gentilis*, they have illustrated it with a photograph of an African species, known as a Chanting Goshawk. The following day they publish an amusing riposte:

Conor Jameson has clearly been looking in the wrong place all these years. There are a number of us thriving in Essex.

Stewart, Helen and Andrew Goshawk, Billericay

As it happens I know Stewart Goshawk, as he manages the environment programme of the City Bridge Trust. The charity has supported RSPB conservation projects for many years. We have a chuckle about *The Guardian* letter when I run into him at Rainham Marshes nature reserve on the eastern fringes of London a short while later, and I make sure to get a photograph of him (complete with name badge) so I can send it to a few people, captioned, "at last I've found – and photographed – a Goshawk ... at Rainham, of all places ..."

I've been planning to visit a friend in Prague, another reputed Goshawk urban stronghold, hoping to find them there. To accommodate another meeting I put the flights back a day. As fate would have it the flight I let go is the last one cleared to leave before the ash cloud from an Icelandic volcano takes all aeroplanes out of the sky for several days. At least it means I am here when another Goshawk is reported on a local website.

"Hi all, what a day to skive off work! After receiving a text from Dean on the 17th with news of his GOSHAWK, I have just totally jammed the same female circling high east this morning being mobbed by a Buzzard! I noted a large *Accipiter* in the same area about two months ago (I did mention it), but views were insufficient to claim anything."

Saturday

Another message comes through. A Goshawk has been seen – and now filmed – over a small village not far north of here. It could be the same female reported a few days ago. This is the verbatim account:

"A pretty ordinary morning was greatly enlivened by the female Goshawk. It caught my eye drifting south towards the village at 14:35, mobbed by a male sparrowhawk and a handful of jackdaws, before I lost it behind the small wood near the car park. I quickly drove to the south end of the village by the farm and soon picked it up again over where it was bombarded by the local rooks. It then drifted back towards me and showed fairly well where I was able to obtain a short video clip! The bird was then on view (on and off) for around the next hour (hence others saw it) occasionally seen in brief confrontation with some local buzzards. Last seen coasting southwards in the distance."

Southwards would bring it towards our village ...

Chapter 6

ON THE ROAD

MAY

Investigation of the local report will have to wait. I'm making last-minute preparations for getting on the road. I've set books out on a table in the back garden in the sunshine, under a confetti of newly opened cherry blossom. I planted this tree a decade ago, and it is repaying me with a great dome of translucent white petals that stretch the width of the garden. I have become more alert than ever for Goshawk. Never has the possibility of it in the sky over the village seemed so plausible. After all, it was there, it was filmed, just a few miles to the north, a few minutes' hawk flying time from here.

I check the sky again. Hazy sun, and a breeze ruffles the map on my table under the parasol. Woodpigeons watch me from the fence, Blue Tits pop into the brick pillar nest hole, and Goldfinches squabble over niger seed next door. The alarm goes up from a Great Tit, the sparrows and co hit the hedge and wriggle into it. I almost flinch myself, and look round to see a Sparrowhawk traversing the sky, some distance away. I'm always impressed that the little birds can see them from such a distance.

Around midday I notice another hawk drifting across the sky to the north, about 100 feet up. I jog to the bottom of the garden to see more. It gives a brief beat of wings, shallow and quite fast. It is big enough to be a female Sparrowhawk, which is nearly big enough to be a small male Goshawk. Female Sparrowhawks, if breeding, should be on eggs by now, mostly. Male Gos would be out hunting, provisioning a nest. Again, if breeding ...

It goes into a diagonal stoop, wings in, at the farm buildings beyond the pasture, and disappears. I wait for a few minutes, to see if the bird will re-emerge, but if it has, I can't see it.

From bed I can often see a Sparrowhawk circling overhead, on reconnaissance. Lately I've seen a female use the line of the road and the houses as a regular beat. And on another recent morning a male made an unsuccessful and slightly cack-handed attempt to extract a Dunnock from the hawthorn hedge in the garden below me. He sat for a while on the hedge, eyes bulging, contemplating his next move, while the sparrows buried themselves in the thicket of thorns nearby. I often sense at times like this just how desperate Sparrowhawks must become, to find something to eat. How quickly their hunger must turn to starvation, and their judgment become impaired, their sight and flight muscles weakened, their risk-taking extreme.

I appreciate such moments to admire the beady eye and striking attire of the hawk, all spindly, mantis legs and bold barring. I wonder how this fragile, nippy little sparrow-snatcher could be mistaken for a Goshawk, and what a Gos might look like down there, in its place. Unlikely to be wasting much time on a Dunnock, or trying to find a way in to a tight little hedgerow like this one, I'm thinking.

Thursday

I have a month of leave and my mind is set on the Goshawk. This is my chance to get out there and find it. I have been planning a Goshawk adventure to far-flung Goshawk haunts, a round-Britain tour to visit Goshawk places and Goshawk people. I load the mountain bike into the back of the car, with bags and notebooks and boots and provisions, and ease on to the great north road again at noon on an overcast day. Minutes later I am passing the scene of the local Goshawk report. After an hour and a half I stop for a quick look around Sherwood Forest. If Robin Hood could survive and thrive here, so should Gos. I find only Robin, in the guise of Russell Crowe on a billboard, and a fridge magnet Fox in the visitor centre.

In an 1869 book called *The Birds of Sherwood Forest*, W. J. Sterland writes that "the Goshawk (*F. Palumbarius*), rare in Scotland, though said to be resident there, is still rarer in England. I never saw the bird on the wing, and only once in the flesh, and we seem to know very little of its life history. Rare as it is, a single specimen was killed by one of the keepers near Rufford in 1848, being the only instance I have known of its occurrence, and I am thus able to add it to my list."

An hour further north, I reach Lamberton, to stay overnight with friends. I want to see the stuffed Goshawk again. Elvis is still here, on the door, to greet me. There is just enough time to look again for the Goshawk, on the off-chance it might still be in the junk shop, before it closes. It hadn't been for sale, after all. I note other taxidermy exhibits, mostly above head height: a Roe Deer head, so mangy it barely resembles the living animal at all, more like something dug from a grave. Ghoulish. A Snipe in a case has the notice "Do not feed the duck" stuck to it. A mounted Badger head comes complete with spectacles. There's an almost colourless, dusty Golden Oriole and a moth-eaten Mink among the other exhibits. But no Goshawk. The Goshawk has gone, who knows where.

"We're closing now," barks the man who's been watching me.

Friday

You have to be up early to catch a Goshawk. I head north again first thing, without disturbing my hosts. The North York Moors are hiding in raincloud;

a waste of a rare, free May morning, that precious commodity. This month can be sublime, and it can be mean with its favours.

I take a left after crossing the River Tyne at Newcastle, and head into the Cheviot Hills. Then it's left again following signs for Kielder Water and Castle, leading eventually to the valley of the upper Tyne. In heading north I am rewinding spring: buds get smaller, daffodils fresher, Swallows fewer.

I'm homing in again on Kielder, the place that has been one of the strongholds for the Goshawk in its struggle to recolonise our islands, a wide forested upland area of Northumbria, close to Emperor Hadrian's great Roman wall, mirror roughly the present-day border between Scotland and England, from the North Sea in the east to the Irish Sea and Atlantic beyond, in the west. Those huge conifer trees loom large and dark as I approach.

Tree planting began here in the late 1920s. The old private estates needed to boost income after the First World War, and forestry was one of the things government would pay for. As plantations mature, the gaps close, light is excluded, ground flora withers, soil is sucked dry and birds and other life forms can't easily move within the plantation unless it is carefully thinned. These plantations can become a refuge for the Goshawk, when they've been in place for a few decades. Looking for the bird has given me a fresh perspective on these forests: something of interest – intrigue, even – to look for here.

The particular state-owned forest I am approaching is a place that has proved secure for the Goshawk since around the time I was born, and during which much of the UK-based study of the species in the wild has been carried out. I pass encouraging signs for Hawk Hope on the edge of the huge reservoir that winds through the planted hillsides around it. Today the water shimmers steely-grey like the sky. There are conifers of varying age and stature, but most are now maturing. I have been to this part several times now, looking from the outside in, but today at last I hope I'll see it from the inside, guided by someone who knows it well.

I draw up outside a small, grey-stone castle on a hillock, some mature broadleaf trees standing guard around it, protecting it from the besieging army of plantation conifers beyond. It was built in 1775 as a hunting lodge. The Arthurian scene is shrouded in low mist. A Cuckoo flies across the set, flickering like a falcon, sending a murmur through the smaller birds foraging in the tree-tops. The man I am to meet operates from the Castle. His loud, sonorous voice is issuing from through the back, behind the visitor reception, when I enter.

He comes off the phone and through to meet me. His name is Brian Little, and he could easily pass for Merlin the wizard – white-haired and bearded, open-necked shirt revealing a pendant on a leather cord. His feet are almost bare. We go to a large room upstairs in the castle, part office, part store-room for conservation hardware. Next door houses an interpretation facility. It has a replica Osprey nesting platform as centrepiece, below a screen beaming live pictures from the nearby Osprey nest. A female Osprey sits there proudly, on top of the world, looking strangely self-conscious and alert for a bird in such a lofty position; as though she hasn't quite adjusted to the celebrity. The Ospreys are spreading south from their Highland stronghold, to which they returned in the early 1950s. Now they have reached England again. This is the second year they've been at Kielder, but the first year they've been fitted with a closed-circuit camera. Ospreys are box office. The centre here is expecting a surge in visitor numbers as a result.

A booth in a darkened room provides recorded footage of Goshawks in action. Old footage, I think, maybe borrowed from a film partly shot here almost 20 years ago by Hugh Miles and screened on national television two or three times. *The Phantom of the Forest*, they called it. The hunting sequences in that film use captive birds, and show a Gos intercepting a crow in mid-air, and taking it to ground, grasping it in one foot, while appearing to anchor itself with its free foot. The crow dies quickly in that lethal, kneading grip. There are nest shots too, and a sequence showing young fledged hawks scrumming over a food item left for them in the forest.

Unlike in the New Forest, no camera has been installed at a Goshawk nest. That is difficult to organise. Goshawks are publicity-shy compared to

Ospreys. They nest in less prominent, glamorous and well-lit locations, and they tend to use the back door for their comings and goings. They are also less predictable in their choice of nest site, and extremely alert to unusual bits of hardware appearing nearby. A colleague in Wales has told me of a scheme a few years ago when a camera was installed at a new Goshawk nest, and it had to be removed again when it became apparent the birds weren't going to come back while it remained. Camera removed, the birds resumed their duties. The same man told me he'd been watching a flock of Red Kites at a feeding station near there, when a Goshawk hove into view, went into attack mode, and took one of the circling kites by the wing. The pair plunged earthward, separating just before they hit the trees below. He thinks it was a juvenile Gos, flexing its muscles.

I settle back to listen to Brian's story. He speaks with the air of an old sea dog recounting tales of the high seas: bright, alert eyes, and occasional flashes of ire and passion as he recalls incidents, and speaks of the birds he has been working with for 50 years. At 74 he is as old as a lot of the trees here, many of which have reached or passed the age at which they can be harvested for timber and paper mills. Planting began back in 1928 and took many years. A lot of the work was done by women.

Tree felling now takes place year-round. As he is no longer as mobile as he'd like to be, the job of monitoring the Goshawks here is done by others, employed by the state forestry agency to advise on felling locations and timings, to minimise or avoid completely any disruption to Goshawk breeding. Before this system was formalised, it wasn't unknown for Goshawk nests to be accidentally removed. When this happened, young birds would be put in baskets and set up on platforms raised as close as possible to the fallen nest. If this is done quickly, the adults will usually carry on rearing their brood. The bond to well-grown young is strong in birds, as a rule. They won't abandon easily – even the highly-strung, capricious, occasionally sociopathic Goshawk.

We are in the middle of the biggest forested area in England – 155,000 acres, or 600 square kilometres according to the brochure, or 25 by 25 miles of plantation blocks. Kielder Water has 27 miles of shoreline to explore. These are impressive dimensions for our crowded island. You can pay a nominal toll fee to take your vehicle (at maximum speed 15 mph) on something called the Forest Drive. This unmetalled track runs for 12 miles to the east, and is one of England's highest roads, they say, reaching 450 metres altitude in places. The forest is also nowadays traced with mountain bike and orienteering trails. Northumbria is England's least populated and most peaceful county, they claim, a tranquillity sometimes broken by the military training ranges on the open ground to the east.

There isn't much for dispersing Goshawks here. What sheltering woods there might be are small. The brochures promise the visitor only Buzzards and Red Squirrels in the trees. The bare hills have been grazed clear by hardy breeds of sheep for 600 years, helped in this by feral goats. The moor mat grass turns to a straw colour in winter, hence the local name White Hills. Eagles have been trying to survive here, but it's too exposed to support a visiting Gos for very long.

Brian has been part of a group of academics and enthusiasts monitoring the population of Goshawks. Numbers have inched slowly upwards to around 28 breeding pairs. Each year, around 50 young Goshawks fledge from these nests and move off to make their way in the world. Around 1,500 Goshawks have been fitted with numbered leg rings in the course of the painstaking, long-term study in the Forest. The idea is that if and when these ringed birds are recovered, alive, injured, dead or accidentally trapped, then a note of the leg ring number will tell where the bird has come from and how far it has travelled. Mysteriously, only a tiny percentage – around three – are ever found. The rest of the ringed birds simply disappear.

All that effort, All that passion. All that potential knowledge of how the Goshawk might try to recolonise our Goshawk-free ecosystems. All that bird.

How can so few of these very large and eye-catching birds, with their thick legs and impressive talons and conspicuous leg rings, have caught the public eye? Alive and well and lurking in the forest somewhere, yes, they remain unseen. But dead too? Proportionally, more leg rings are recovered from the Willow Warbler, a species so tiny it would fit in a breast pocket with barely a bulge.

It doesn't stack up. Brian tells of a study done of Goosanders, a river-dwelling, fish-eating duck, for which licenses can be issued to control numbers of the birds, on the grounds that they eat too many fish. Loads of rings are recovered from these birds. They are shot – legally – and the rings are duly handed over by those who have shot them. But not Goshawk, which of course cannot be killed legally.

There has been an exception to this overall lack of Goshawk retrievals: a juvenile bird, in its first year, recovered in the Lake District, a hundred miles away to the south-west, where it had travelled to spend the winter. This recovery record show that they can disperse long distances in winter, to find space and food. After that, if allowed, they will tend to drift back in spring, closer to where they were raised.

Brian describes the first Goshawks here, which appeared and attempts to breed back in 1974. Someone tried to make them desert the nest, by hanging around the nest tree for a day to keep the birds off the eggs. Brian talks about the female Golden Eagle that was tracked across Scotland using transmitters. She travelled the length of the country, from the Border country of the south to Perthshire heading north-east, up north and over to Orkney. She spent the night there and flew back. She made light of these distances. Eagles can reach 70mph in level flight. At altitude, they can drift for miles with minimal effort. We think of them as wide-ranging, but confined to certain areas. In fact the whole country can be their domain.

She died after finding a dead grouse, laced with poison. In some ways it was fortunate that she ate some of it and died just before she reached her nest, and her only chick, with the lethal bundle. She died below the eyrie, which allowed the chick to be fostered, and to survive. Brian becomes heated when talking about this kind of illegal destruction of birds of prey. He believes that young Goshawks might just about manage to survive a year beyond the forest here, even nesting once, as they are secretive, but will rarely be allowed a second year before they are removed.

"It's the prerogative of every member of the public to be able to see these birds of prey in the wild," he says. I can't really report what he said he would like to do to anyone caught killing them.

Brian takes me to the forest, me carrying his ladder that we fetch from the castle's cellars. But not to Goshawk nests. Brian has restricted mobility now and the Gos nests are too deep in the trees, as a rule. Birds are on eggs just now, not doing much else. The raptors that rise above the conifer tips on the skyline are Buzzards. Instead we visit his Tawny Owl nest boxes. More than a hundred are occupied this year.

Brian is pleased because he has proved conclusively, by analysing prey remains in owl boxes, that Kielder's Tawny Owls, especially in years when the Field Voles that live in grassy areas are less abundant, will predate even female Sparrowhawks.

Goshawks sometimes catch Tawny Owls, as well as Sparrowhawks. And by day, Sparrowhawks will take young owls that have ventured out onto branches to rest and roost. It's a bird-eat-bird world.

As we leave the forest, a large bird rises in the distance. A Buzzard is mobbing it. I point it out to Brian. "There's nowt wrong wi' your eyes," he laughs, in his broad Geordie accent. He stops the car to look more closely. Languid, bow-winged, it was making the Buzzard look small. Osprey. Apparently there are three here just now: the nesting pair, and a third bird, probably a spare female, hanging round and being largely tolerated. It was a fitting and pleasing way to finish the day, and our chat. No Goshawks, as such, but already I know much better the species I'm dealing with, and some of the issues it faces. This trip isn't all about seeing one, as I will find myself repeating. It is about getting to know it better, and how to help its cause.

As well as young Goshawks dispersing from Kielder to take their chances in the wider world out there, a few Goshawk disciples have cut their teeth in these forests, learning about the birds here, helping with the long-term study, taking what they know to other parts of the country. I have lined some of them up to find out whether the Goshawk has gone with them, to other forested areas of Scotland, further north again.

Friday

Mum and Dad retired some years ago to the Borders region of south Scotland, not far from Kielder, on the Scottish side of the border, which runs nowadays north of Hadrian's Wall. They are tucked away in a quiet corner, conifers and oaks towering over their bungalow. It is early evening by the time I get there.

My next Gos appointment is an early morning rendezvous with Malcolm Henderson. I haven't seen him for almost a decade, and that first attempt to find the Goshawk in its secret world here. He is now retired from the police, but still going strong with the raptor monitoring. He has brought his young 'protégé' Caroline Blackie, who helps with some of the work, has a lot of expertise of her own, and can carry this on into the future, all being well.

We will be going to visit some Goshawk sites, including birds that may have dispersed from Kielder. Malcolm keeps tabs on around 20 Goshawk nest sites across his Borders study area.

Langholm Moor. We meet conservation staff in a large farmhouse nearby. Wheatears are arriving. A male stands proud on the tall stump of a formerly massive tree, now removed. Staff here are involved in the project that is monitoring Hen Harriers and other wildlife on the moor, alongside the populations of Red Grouse that are managed here for shooting. It's a long-term project to explore ways of balancing birds of prey with viable grouse shooting. At present there is one pair of harriers locally. We don't see them, but we see a Buzzard hunting almost in the style of a harrier, quartering low along a burn within a marshy area. We also see a Merlin, tiny by comparison, mobbing the larger raptor. We chat to a friendly gamekeeper, a real nature enthusiast, genuinely sympathetic to raptors. There's an amnesty on at Langholm, to establish just how things balance out on the predator/prey/shooting front. And a Red Grouse sits quietly, oddly exposed in the low heather, not far from where we are parked, as though secure in our company, like us scanning the wide expanse of moor for any other signs of life.

There is none of the extensive conifer forestry here. It is rolling country, farmed for arable and livestock even on the higher elevated ground. It is studded with woodlands and scattered trees. We park on a leafy track, and walk some distance through mixed mature woodland until we enter a much denser plantation – too tight for Goshawk – the kind of density that Sparrowhawks might use for nesting with the Goshawk around. We reach a more open plantation, with widely spaced, tall, hefty trunks, plenty of light reaching the ground. The forest floor is mossy, with ferns and wildflowers, upturned stumps, stones. Malcolm describes some of the field signs to look out for, and Caroline is already off on a search. She is being informally trained, so like me she'll be given a few clues but have to find the evidence, and the nests, for herself. Malcolm plays a few artful tricks to try to throw her off the scent.

While Caroline is off searching Malcolm tells me about a prank he played on a government official he was training in the art of understanding the Goshawk. It involves him kneeling down beside a splash of white raptor excreta. "If in doubt whether it's Gos or Buzzard," he told his young apprentice, dipping his finger in the fresh mute, "bear in mind that Gos is much more bitter to the taste." He then touched his finger to his tongue, and made savouring noises, smacking his lips, like a TV chef. "This is Buzzard, you see," he decides, gesturing for the trainee to try it for himself. The trainee, eager to learn, and probably wondering how else he would ever know one kind of raptor shit from another, bent down to dip his own finger in the liquid. Malcolm must have been feeling compassionate that day, as he intervened before the tyro went any further. Malcolm had of course switched fingers before tasting. Luckily for Caroline – and me – he doesn't try this on us.

Caroline is very keen to show she can find the nest site, and the tracks that lead us there. Malcolm has visited this site earlier in the season, and already has a fair idea where the birds have chosen for their nest. We'll get confirmation today if they are actually in residence.

Shafts and spotlights of glittering sunshine stripe and splash the lime green mossy rides, and illuminate the whole scene. Malcolm gestures quietly when he finds something. There are prey remains – nothing too obviously fresh, but visible when you know what to look for, and the likeliest places to look for it. The male Goshawk provisions the female while she looks after the nest and eggs. He has particular routines when it comes to bringing and plucking food, the sort of places he feels comfortable doing this, and that will make his approach to the nest straightforward, and his getaways. "He doesn't want to get under her feet …" as Malcolm memorably puts it.

Because females are inactive in this earlier part of the cycle, they moult their feathers at this time. They shed their major wing feathers – or primaries – in a particular order, and these re-grow over the season. We find droppings – the polite name is white splash – pellets, feathers of pigeon, Pheasant, Woodcock, woodpecker, thrush and crow.

We find a fallen Goshawk nest from a previous year, like a mound of kindling. Malcolm rakes in it for any traces of prey. He has given Caroline some red herring clues and she is off searching in the wrong part of the wood. He is testing her. She finds clues anyway. He indicates quietly to me the direction of the current nest, or the likeliest candidate – they can build a new nest very quickly, if for some reason the pair are unhappy with the nest first built. They can also use a previous nest. You will sometimes find several nests in quite a small area. Knowing which one, if any, is in use requires an expert eye, and a bit of intuition. Active nests just look a bit more alive, if a mound of sticks and twigs could be said to have life, viewed from 50 feet below. It is impressive to see such expertise in action, such a trained eye at work.

Caroline may have overshot the nest site, but she has seen the female Goshawk leaving the scene. I didn't see or hear a thing. The bird doesn't call as we quickly take our notes and leave the area, not wishing to keep her away from the nest for long.

The forest is quiet, save the sighing wind deflecting gently away off a billion needles high above. All within and around the columnar, flaking trunks is still, calm. The only birds apparent are a couple of Wrens furtive in the brash, and more Crossbills moving through the pine canopy, revellers on a pine cone binge. Here be Goshawks, and Goshawk nests, and – all being well – the scene of weeks of Goshawk activity. But almost none of it will be witnessed, and certainly none of it by me.

Malcolm knows of one Goshawk nest site close to a rookery. The remnants of around 20 Rooks were found at or below the Goshawk nest at the end of the breeding season. Goshawks will often hunt early in the day, and take a smash-and-grab approach to rookeries, lifting young birds from on or near nests, amid inevitable pandemonium. It's quite a difficult thing to witness, when the spring is advanced and the tree-tops in full leaf. It all happens very quickly too, of course.

Malcolm is apologetic again that we haven't seen any Goshawks. I remind him that this isn't why I'm here.

Thursday

Aberfoyle. This is heavily forested country. A large part of it is constituted by the Queen Elizabeth Forest Park, managed again by the Forestry Commission. Based on the evidence of other forests I've seen so far, it should be Goshawk country. But strangely enough, it isn't Goshawk country. Designated in 1953 to mark the Queen's coronation, it extends to the eastern shore of Loch Lomond, across 50,000 acres. The website promises the following highlights: "Ospreys are the major draw, and they have bred here since 2004. Red squirrels still thrive here, and red deer, wild cats, polecats, pine martens and water vole are all present. "Apart from the ospreys," it goes on, "the reserve is especially well known for birds of prey. Golden eagles, sparrowhawks, kestrels, peregrines, buzzards, merlin, red kite and hen harrier can all be seen." There is no mention of Goshawk.

Dave Anderson has studied Goshawks in different places for many years, and knows these forests intimately. After a chat over a cup of tea, we set off to visit a couple of sites, with Hamish the terrier in the back of the truck.

The scenery here is dramatic – besides the maturing forestry plantations there are open hills, lochs, broadleaved woods, small fields and rivers. Roads wind up and over the sage-green glens past delicately feathered loch-side willows, and glittering wavelets. I open the gates as we head up steep tracks and hairpins into forested areas. The trees seem especially imposing on slopes, where they tend to grow even taller. A Red Squirrel poses for us at the base of one of these 80-year old Norway Spruces. It is tiny compared to its grey, imported cousin. Its ears are tufty, tail blond, movements light and prancing. It conveys no impression of weight. Not much of a meal for a large raptor, say. Grey Squirrels may look quite small, but they weigh three times as much as Reds. My notes from this forest form the prologue to this story.

Dave has noted from other places where he has studied the Goshawk that they seemed to submit to Buzzards when it comes to competition for nest sites. This feels odd, considering the Goshawk is a far stronger, more rapacious species, and will include Buzzard on its menu when the occasion merits it. But Buzzards are noisy, obstreperous neighbours. Goshawks often seem to just want a quiet life, to breed in peace, where possible. The presence of Buzzards will sometimes make the Goshawk go into what seems like less ideal, tighter cover, a less mature stand of conifers, in some cases, with less space between, less height, less robust structure. They seem to do OK in these places, however.

Among the many unique insights his many hours and days of looking out for the Goshawk have given him, Dave has witnessed adult Goshawks deliver live pigeon prey to their young, be believes to give the young practice and experience in chasing and catching food.

He knows of one apparently reliable record that a Goshawk killed three Pheasants in a pen on one visit. It is impossible to know how many Pheasant pens there are. Much of it is done quietly and privately, and not declared as income. An estimated 70 per cent of Goshawks are lost, missing in action, unaccounted for, judging by ringing data.

We discuss the curious lack of recovered rings. Dave did get one back, once, given to him anonymously by a keeper who had shot the bird. Dave explains that there have been Goshawks here in the past, and that they have bred, relatively secure within the forest in the breeding season. The problems arise when they disperse. Adults and young tend to wander outside the breeding season, moving after the thrush species that arrive in autumn and winter from northern Europe, and which feed on berry crops of species like Rowan and hawthorns as they move south through Britain. Goshawks follow these, and Woodpigeons. And then they vanish.

***

I spend the evening as a guest of Patrick and Susan Stirling-Aird, who live in a beautiful location in the woods near Stirling, by the banks of the Allan Water river, a place I knew very well as a student, and for a couple of years after graduating. It is a setting strongly linked in my mind with Robert Louis Stevenson, who spent time here as a boy and wrote fondly of it. Stevenson's Cave is ensconced among trees by the river, and is thought to have fed his imagination and featured in another guise in *Treasure Island*, as well as his tales of the Jacobite risings.

The log-burning stove hisses in the late evening, as I discuss birds of prey with Patrick. He is Secretary of the Scottish Raptor Study Groups, and a lawyer by profession. Opponents of bird of prey conservation have been lobbying for laws to be weakened, so that birds of prey can be legally killed. It is already theoretically possible, under wildlife law, to obtain a licence to control Buzzards. A strong case has to be made that there is a justification, that a Buzzard is causing enough of a nuisance, and that there are no non-lethal solutions, before such a licence would be granted. No licences have been issued so far. It's hard to imagine how removal of a relatively scarce native species like the Buzzard could be justified in the context of the supernormal densities of non-native game birds that prevail in most places.

Some shooting interests have tried to challenge the status of the Goshawk as a native species. They would like us to believe that it never was, and that the historical records can all be put down to imported falconers' birds. There is overwhelming evidence to the contrary, but it is one more debate that has to be aired in the corridors of power.

Monday

At breakfast, conversation turns to the unusual bird of prey that has been seen locally, including perched on the wall of the Victorian kitchen garden, which is flanked on its fourth side by the river. The mystery raptor has already had one of their chickens, and was discovered there one evening with its catch. My first thought is that this might be a Goshawk, but Patrick of course knows his raptors, and the bird he describes, that he has seen well and unusually close, as it watched him from on top of the old cottage roof, he thinks must be an escaped Peregrine/Gyr Falcon cross. Hybrids such as these can be kept by anyone, without a license. An inexperienced owner may soon discover that he or she lacks the time, energy and expertise to keep a raptor like this, and it soon escapes, or flies off, or might even be let go.

I explore along the riverbank in the early morning. There are animal footprints in the sand by the water's edge – possibly young Otter, if not Mink. A Grey Heron flies languidly upriver, and not so languidly when an irate Buzzard flashes out of the trees and spooks it into a surprisingly deft evasion manoeuvre. The old gardener's cottage stands as a ruin above the river, looking through eyeless windows at the shallows below. It is said that the gardener's family died of typhus. The fireplace is built around a window from Dunblane Cathedral, which had fallen into dereliction by the early 20th century. Because the estate-owning family here donated to its restoration fund, this piece from the old Cathedral was given to them as a gift.

I was christened as an infant in the restored Cathedral, so this all feels very close to home. I lived a little way down river from here when I was a student, and for a happy summer after graduating. It was a house in the woods, on the edge of the moors. From my bed one morning I caught sight of a mystery raptor soaring over the house.

My student diary from 1986 has this note for Wednesday 15 January:

The westerly gloom had gone and the sky was clear. The wind though was biting as I set out on a walk ... I had seen, from my bed, a large raptor drift over the house towards the woods. Buzzard? It looked more like a harrier – goshawk crossed my mind. I craned my neck out and saw a definite buzzard soaring beside some gulls. I couldn't believe it. Shouting to Jerry I ran out in my pants and saw what looked like the first bird above the woods behind the house ...

Later, on the walk (housemate Jerry stayed in bed):

I saw two small hawks in a high-speed dash across the end of a field closely pursued by some wind-assisted crows ... the large raptor reappeared, heading nonchalantly across the sun in the direction of the far highlands so clearly visible and distinct although many miles away.

Buzzards were just coming back, then. I still wonder though if the first bird I saw that day might have been a wandering, first-year Goshawk.

After breakfast I head north towards those Highlands, across Perthshire and into the Cairngorm National Park, with its Victorian Heritage routes signposted as alternatives to the main drag. I take these, and they lead me onto quiet roads over bleak expanses of moorland. Pockets of snow and flurries of sleet and hail add spice to an otherwise bright sunny day. Eagle country. After crossing the high ground, I descend through Caledonian (Scots) pine forest, elegant and historic, open woodland, with a healthy under-storey of heather and blaeberry, into the heart of the country, a place often recording the coldest temperatures in Britain in winter. The thermometer readings are certainly plunging today. I make my way along the River Dee's valley, to Royal Deeside, past Balmoral on the right and Crathie Kirk on the left, where the Royal Family attends services. Goshawk country, one would suppose. I wonder how Goshawks do in these old Scots Pine woods, of the kind once extensive in the higher ground and the north of Britain, before being cleared through the centuries, and in many cases replaced with exotic firs, farmland, grazing and grouse moor.

I meet Mick Marquiss at mid-day, and it is snowing by the time we reach our first Goshawk territory, deep in commercial forest, on high ground. To answer my first question, he tells me he's found just one nest territory in old forest. We come to a halt beside a large clear-felled area of a modern age plantation. There was a Goshawk nesting site here, he explains, but the nests failed for three years in a row, with evidence of the female being shot in at least one of those years. Clear-felling would at least encourage the birds to try elsewhere. Mick has recently seen a pair, and is hopeful the male has found a new female. He also has a fair idea where they have moved to, based on visits he made earlier in the season, at the stage when the birds establish pairs, territories and nest sites. We head there next to try to confirm that a nest has been set up and is occupied.

Sheltering from the hail, I can see across to open moorland. The terrain here is generally drier, rockier, less mossy than in central Scotland, where it rains so much more from Atlantic cloud. Before we enter the wood Mick explains how we should walk a certain distance apart, not speak in the vicinity of the nest, and signal silently if we find anything by way of evidence of birds, or anything suspicious. The forest rises on a fairly steep slope and some clambering is involved in getting across fallen branches and trunks, from forestry thinning operations, and snow-broken limbs brought down by this winter's heavy blizzards. Rowan trees seem to have been especially brittle, partly because they stretch so hard to reach light in competition with the softwood firs that they become top heavy, shallow-rooted, and topple easily when pushed by a gale. The tracks are heavily churned by mechanical shearing machines. We circle high up then back down in the direction Mick thinks a new nest might be.

He gestures to me that he's found something. "Grouse remains," he whispers, when I get there. There are feathers, and bones, and even grouse feet. Who says these are lucky? I'm sure they still sell them as brooches in tourist gift shops, one by-product of the shooting business at least. Mick points out something he calls "snow shite" – bird droppings that have fallen on snow, which has then melted to place them gently onto the ground below, to stand proud on the moss, dried bracken, twigs and needles of the forest floor. His eye for detail is forensic. We find larger bones. Mountain Hare. An experienced female Goshawk is capable of catching and subduing an adult hare, but these are heavy animals, with a powerful kick in their hind legs. It is mostly young hares that they take, lifted back from open ground to the forest, giving some idea of the muscularity of these hawks. The male bird is doing most of this fetching and carrying at this time. He is a good deal smaller than his mate, but still a significant and muscular predator. We find Golden Plover feathers, Mistle Thrush. Mick is able to identify immediately from the remnants not just species, but age, sex and condition of the birds. He knows the old stuff from the recent, and he records prey items to build a picture of what Goshawks are eating, and when.

We are also looking for Goshawk feathers, partly to confirm they are in residence, but also to identify individual birds. Mick can sometimes do this from the feathers themselves, by eye, based on others he's collected. Failing that he can have the DNA looked at in a laboratory, but this costs money, which soon adds up when you are covering a study area as large as his, here in north-east Scotland.

We finally reach the site of the nest. In a surprisingly flimsy larch tree high above us hangs a peculiarly two-tiered nest. It is one old nest, which has partially collapsed under weight of snow, and a new nest just above it. A branch is shattered where the earlier nest has come down on it. Snow snap, Mick calls it. And we find something odd: the fragments of a smashed egg among the tangle of nest material that has reached as far as the ground. Mick gathers this for examination later, but it looks like a Goshawk egg. He isn't too concerned, because by all other appearances this is an active nest, the birds are in residence, and although we haven't seen or heard any today, that will change. We don't hang around. As ever, we need to minimise the time that the parent birds are away from the eggs.

Goshawks can build a nest in a matter of days, where necessary, like when they lose a nest or have to replace one, or move to another site for any reason. The male starts it off, then the female takes over. They use mainly branches from the immediately surrounding canopy, and some from further afield, breaking them off with feet and beak, and interlocking them. Larch is often favoured, as it is easy to break, pliable, sinewy wood, and the 'knobbles' mean it holds fast.

Mick is pleased like me to have had a real winter this time round. Snowfields are back on the mountains. We can see pockets of white on the hills beyond. There are little clusters of hailstones at the base of trees, out of reach of the thawing breeze, which is just above freezing. A Rabbit lies dead in a thicket, perhaps starved and frozen to death in late winter. Vole tracks riddle patches of the forest floor, where they have been tunnelling and nibbling under snow. Patches of wind-thrown trees, knocked over like skittles and leaning against the ranks behind, sometimes allow hawks access points to dense woodland. They like a clear flight-path, especially when carrying heavy cargoes. Goshawks like to fly in to the nest from below the canopy. Buzzards like to access from above as well; they are skylight birds, preferring to maintain height, and not usually carrying such heavy prey. If Goshawks are around, Buzzards probably also like to know what's above them.

There are droppings, like dabs of greasepaint flicked from a brush; pellets of fur and feather and bone, and tufts of squirrels and Rabbit fur on tree root stumps, all draped in lichen. Sometimes these plucking posts are muddied by the shuffling feet of the birds – heavy talons, manoeuvring prey into position. There is plenty of evidence of Foxes here, conspicuous droppings that mark their territories and leave messages for others. Pine Martens live here too. They will sometimes fight with Goshawks at the nest, and may occasionally take Goshawk young.

As we leave the forest we hear first the male Goshawk calling, and then the female. She has the deeper voice. It is an almost chimp-like chatter, today it sounds mocking. We can see you, she seems to be saying. But you won't see us. Oh, and by the way: don't rush back ...

There is something about this male bird that keeps it safe from crow traps. It just won't go in them while almost all Goshawks cannot resist, and the free meal they seem to be offering. These traps are large crates, with a decoy crow inside. This is legally set to lure other crows in. It is permissible under licence to destroy crows under certain conditions. The birds can get in, through a lobster pot style entrance, but cannot get out. They are magnets for Goshawk, perceiving a free meal of crow. If a Gos is caught, the trapper is supposed to let the bird go. It's a matter of trust. But no one is looking. Usually.

A Roe Deer vaults through the plantation below us and across the storm ditch and cleared areas on the edge of the wood, its white bum like a bobbing face amid the darkened trunks.

Mick has been following the fortunes of the Goshawk here since 1973. Back then, he knew a gamekeeper who asked him one day: 'Mick, what sort of hawk can lift a Pheasant out of a pen with one foot?' Mick already had a fair idea what species he might be talking about, and he knew thereafter to be on the alert. (Mick's story reminds me of the time, a few years ago, when I met the manager of a lime quarry near Cambridge. As we stood on the rim of this vast canyon, gouged over years out of the chalk ridge, he described to me having seen a bird of prey catching a Pheasant way down at the bottom of the great hole, then lifting it out of there. At the time I was perplexed, thinking only an escaped eagle or similar might be capable of such a feat. I now realise it could have been a wandering Gos.)

Years later, the same man, now retired from keepering, confessed to Mick that he had shot that same hawk. And another one, another young bird, that same year. They may well have been the first young Goshawks produced in the region. Mick was studying Sparrowhawks and Buzzards at that time. Sparrowhawks had been in severe decline across Britain through the 1960s, due to agrochemical poisoning. Back then, he knew all the Buzzard territories in the region. There were just eight of them then. Mick recalls the excitement of one day seeing his first live Goshawk in the wild – a juvenile female. Unmistakeable. He knew there must be a nest somewhere, but of course back then no one had any experience of how and where to look for Goshawk nests. He persevered, and went through all his study areas, until finally, late that autumn, he found what had to be the Goshawk nest, his first taste of the subtle, tell-tale characteristics he now knows so intuitively.

Mick is an avowed optimist. He believes the Goshawk is winning, as he puts it. 'They can't kill them all now,' he says. They are getting clear. There's now a pair on the edge of a major city. 'No one really knows they are there. But there's so much food for them in the east here. So much rich soil, fertile and productive, it produces lots of prey in the form of crows, Magpies, pigeons, thrushes, gulls, Pheasants.' It's still a different story in the highlands and the forests to the west, he believes. There is much less food there, even in the native forest. They don't seem to do well: even without illegal killing there are so few Jays and Magpies, and even few Woodpigeons. Certainly nothing like the densities of prey you find in the fertile, fertilised lowlands.

News comes through that Police are investigating the suspicious deaths of three Golden Eagles, a Buzzard and a Sparrowhawk on the Highland estate best known to a wider public as the place where Madonna and Guy Ritchie were married. The birds were found on a Grouse moor north of Inverness last weekend. It's one of the largest single incidents involving the suspected persecution of birds of prey in recent years. It's known that well over a third of Golden Eagle territories are unoccupied in the surrounding region.

Received wisdom is that the Goshawk became extinct in Britain. But can we be sure? It's tempting to wonder if a bird as elusive as the Gos might have hung on somewhere, unseen. The thing that clinches it is the fact that persecution was so widespread, and Gos are so easy to trap, or to lure with the offer of a free meal. If they were here, they would have been found. So, no one seriously disputes that the Goshawk was wiped out in Britain by the Victorian period – the last decade of the 19th century. There were no longer any left here that could pair up and breed. They were gone.

Hawkwatcher author Dick Orton thought he might have worked out when the last one was shot. According to the brief biography on the book's sleeve, Dick gave "over 6,000 hours of voluntary service to guarding the eyries of merlin, hobby and – above all – peregrine falcon." He came from a family of gamekeepers – "all of them men dedicated to the destruction of birds of prey". All, that is, except Dick, who broke the mould. His book is a trove of beautifully described accounts of looking for, and looking out for, rare birds of prey, and guarding their nests.

He's even had some direct experience with Goshawks. He talks about the surviving 'vermin' records left by a gamekeeper at an estate in the north-west Highlands of Scotland; a place called Glengarry. The keeper, whose name was Murdoch Mathieson, recorded shooting a Goshawk. Although Mathieson doesn't give a precise date, Orton speculates that the event happened late in the 19th century, and might therefore be the last record of the Goshawk here before its extirpation.

I wonder if I can find Glengarry, where this final act in the life history of Britain's original Goshawks might have taken place. A morbid pilgrimage, perhaps, but it would be interesting to try to confirm Orton's theory, and maybe even find the Goshawk there, renascent.

Glengarry lies at the western end of Loch Ness, in a quiet and wonderfully scenic part of the north-western Highlands. From the north-east I follow the northern edge of Loch Ness, the great fault line that cleaves Scotland, in the glare of spring sunshine. This is the route taken by Bonnie Prince Charlie as he fled the battlefield at Culloden, after the predictable rout of his tiny, ill-prepared Jacobite army there by government forces. The radio is full of excited, conspiratorial talk of the new coalition government steadily forming in the UK, the unlikely alliance of conservative and liberal. Traffic streams north on the oncoming side of the road.

There's a newly-built heritage centre, with a freshly laid car park, in the centre of a wide strath, off the main road to Skye. The door opens onto a cafeteria, with the sounds and smells of catering activity. Two women are busy in the kitchen beyond a hatch, friendly faces, happy to have this early customer – early in the day, and in the season. They've only been open three days. I tell them about my mission to find a copy of *Place Names of Glengarry and Glenquoich*. I'm assured I will find that here, when the information officer arrives. There's a resource centre I can make use of. I have a cup of tea in the sun. Swallows are racing north upstream and returning Willow Warblers repeat their melancholic, falling trill. An off-beat note is added by a hooting Tawny Owl from across the meadow that borders the narrow river, leading northward into the forest that coats the upper end of the glen. The mountain peak at its head is Ben Tee – the mount of the fairies.

I'm introduced to 'Old Roddie', who is having his lunch, with a can of Irn Bru. We get talking about the birds. He thinks there's "nae much" about here, bird-wise. He laments the loss of Sand Martins, and birds more generally, which he blames on foresters letting the spruces grow where the Martins and other birds used to nest. From where we are sitting I can see a Mistle Thrush taking a beak-full of worms to its nest. A wagtail skitters over the grey shale of the car park.

I tell Roddie about my particular interest in the Goshawk.

"Ah yes," he reflects quite candidly, if a little distantly. "They came back here about ten years ago. I saw one up by the loch, and it went in to the old pine trees there."

The way Roddie recounts this incident makes me think he is right. He has the unforced air of a man who knows what he's talking about, and isn't too much troubled whether I believe him or not. He mentions Buzzard, Peregrine and Sparrowhawk also being about.

"I can't get out much now, myself, as I've a bad hip," Roddie goes on, in sombre tone. "And it's been the hardest winter any of us can remember. Snow on the ground from December to March."

"The scoters are back on the loch, mind you," he adds, briefly cheerful. He also tells me I've just missed the head keeper, who'd been in earlier, whom I could have had a word with.

The information office opens, and I'm ushered through to meet the local historian Stroma Riungu and her volunteer assistant Joane Whitmore. They prove a fertile store of information about the village of Invergarry and the glen above it. I explain to them my interest in the gamekeeper Murdoch Mathieson, and in working out when he was born. Anything that might shed light on Dick Orton's theory.

"One assumes a birth date not much before 1850," thought Dick, "and the death of the goshawk an event not occurring earlier than 1870. If much later (as well it may have been, for Mathieson could still have been young enough for keepering in 1900), his may have been the shot which extinguished the species from the British list for about fifty years. Mathieson certainly described that gos as the only one he had ever seen."

Stroma digs out the census records. In fact Mr Mathieson was nine years old in 1881, giving him a birth year of 1871 or 1872. His occupation is listed as gamekeeper, ten years later, in 1891. Assuming he shot the Goshawk around that time, this would indeed be a very late record of Goshawk in Britain. But would it make his the last bird shot before extinction as a resident species? Unlikely. Mathieson made the point in his notes that this was the first and only Goshawk he had seen or shot at Glengarry, so we can be sure the birds weren't breeding here in that period. His Gos was almost certainly a wanderer, a passage bird.

I do my bit to support the visitor centre by purchasing a copy of the more recent edition of *Place Names*, and some postcards reproduced from drawings taken from the book. Armed with these resources, I head off to explore the glen and the forest by mountain bike. I note a Slow Worm curled up by the side of the track, but a closer look reveals it to be dead, crushed by a tyre, unmarked, but limp, glazed, wasted. Another unfortunate Glengarry statistic.

At the southern end of the loch the Swallows are skimming the wavelets. A sandpiper flicks away from the shoreline, all the way across the loch to the opposite shore. Signs on gates call on us to "Fight for our forests!" with details of a Climate Change Bill website.

I sit on the edge of the track overlooking the forest canopy, mostly larch and spruce tips, with pine and still bare oak, under blue sky, warming sun. I scan the ridge opposite, dotted with a thicket of wind turbines. Some are turning, but most are static. I flick through my new-old book. *Place Names* was written in the late Victorian period, then again in 1930, with the keeper's notes added. I am in the historic domain of the clan McDonnell.

"Beware of McDonnell! Beware of his wrath!
In friendship or foray, oh cross not his path!
He knoweth no bounds to his love or his hate,
And the wind of his claymore is blasting as fate.
Like the hill-cat who springs from her lair in the rock
He leaps on the foe – there is death in the shock.
And the birds of the air shall be gorged with their prey
When the Chief of Glengarry comes down to the fray,
With his war-cry, 'The Rock of the Raven'."

It talks about a place called Badantoig, not far from where I am sitting. It comes from *Bad an-t-seobhaig* – which means grove or clump of trees of the hawk. "A very considerable settlement at one time," according to the book.

"There is not one house left now; but the clump of trees and the hawks are still there." I wonder.

There was, apparently, "no more turbulent glen in the Highlands than Glengarry." It came to be regarded as the centre of all insurrectionary movements in the north, regularly raided by government forces. Cromwell sent a force in 1654 to subdue the clan chiefs, and burnt the house here.

"In the old days there must have been a very large number of birds of prey in Glengarry," says the book, "as appears from the following list of vermin trapped in the glen between 1837 and 1840".

A list of birds and other beasts follows, with the remarkable tally of 63 Goshawks listed as killed. Could some of these have been Peregrines, or other species? There are separate entries for other birds of prey, as follows (my brackets):

27 white-tailed sea eagles

15 golden eagles

18 ospreys

98 blue hawks (Sparrowhawk? Peregrine?)

7 orange-legged falcons (Peregrine?)

11 hobby hawks

275 kites or salmon-tailed gleds

5 marsh harriers

285 common buzzards

371 rough-legged buzzards

3 honey buzzards

462 kestrels

78 merlin hawks

63 hen harriers

6 jer falcons (Gyr Falcon?)

9 ash-coloured hawks or long-tailed blue hawks (male Hen Harrier?)

It is tempting to wonder if the listing is reliable, and the numbers exaggerated, but keepers were usually expected to back up their claims with evidence. Gibbets were checked by the bosses. There was little leeway for keepers to enhance the evidence of their labours, or exaggerate the volume of 'vermin' being and needing to be controlled. I should add that I've not known historians query these points, and they know more about this stuff than I do. The Glengarry vermin list is oft-cited, as it is unusually detailed. Most estates have either lost or never felt the urge to compile and publish these historic keepering records.

Besides the issue of volume, there seems a disproportionate number of Goshawks, and no specific mention of Sparrowhawk or Peregrine. While the gamekeepers of the age were apparently excellent marksmen, spending long hours in the field at the right times of day and season, they were perhaps less inclined to careful and close study of bird behaviour and taxonomy. Could it be that these species were being mixed up and lumped together, perhaps because of the difference in size and colouration between males, females and juveniles of raptor species? Other details seem improbable. Rough-legged Buzzard and Gyr Falcon are occasional winter visitors to the UK, for example. Could they really have been present in such numbers a century ago?

**

In addition to these day-flying birds of prey, there are records from this four-year period of an impressive haul of other species (again, my brackets):

11 foxes
198 wild cats
246 martin cats (Pine Marten?)
106 pole cats
301 stoats and weasels
67 badgers
48 otters

78 house cats going wild

1431 hooded crows

475 ravens

35 horned owls (Long-eared Owl?)

71 fern owls (Short-eared Owl? This name was also used for nightjar in Scotland – according to Lovegrove)

3 golden owls (Barn Owl? Note no mention of Tawny?)

8 magpies

But what of the local population, while all this determined keepering was going on? The book gives a clear impression of how Glengarry, like much of the Highlands more generally, was once well populated, though hard-pressed. "The people of Glengarry were at this time (1841), from all accounts, in a most wretched condition; so much so, that ... out of a population of 315, 35 families, or 122 individuals, were quite destitute."

The clan chief had been forced by insolvency to sell the estate a few years earlier. "Crippled for want of funds, and destitute of men, they were compelled to part with their ancestral home."

Richard Cobden was far from complimentary about what he found. Writing in 1862 of neighbouring Glenquoich, he spoke of "this dreary glen ... which agriculture has abandoned to the dominion of the wild animals of the chase; that a community should be content and happy whilst thus deprived of the benefits of civilisation is a lamentable instance of the triumph of barbarism."

Looking up from the pages of this engrossing book, the only sounds in the glen are a trickling stream, the hum of an insect, and the distant falling notes of another Willow Warbler deep in the forest. Another one answers. A Chaffinch joins in. A Red Admiral butterfly commutes past. I think of old Roddie's sombre admission "there's nae much around". I think he's right. If you accept the domination by 21st century foresters' trees, it is beautiful, but feels eerily empty.

I cycle on, exploring higher into the glen, to ever remoter, quieter forest. I encounter no raptor of any kind, not so much as a Buzzard or Kestrel. I do chance upon a Tree Pipit in full flow, which is gladdening amidst the silence. And following my map carefully I eventually find my way to Ardochy Lodge, where Murdoch Mathieson loosed his two barrels – the first to kill a Blackcap (to prove they could be found here) and the second at the Goshawk that passed overhead shortly after. There were witnesses. I looked at his notebook entry regarding the incident, reproduced here in full:

"The only goshawk I ever saw, I shot a Ardochy Lodge, Glengarry. I was one day at Ardoch and saw a blackcap. As it had been disputed that the blackcap ever visited Glengarry I shot it, and on firing the shot the goshawk came over the Lodge and I shot it too. It fell in the garden. There were several people and children looking on.

"Both birds were sent to Messrs MacLeay, bird stuffers, Inverness, who said that the birds were a blackcap and a goshawk and that both species were young birds. In one of the old game books several goshawks were reported killed. This one, however, must have been a straggler."

I wonder at his reference to "several goshawks" reported in the old game books. What about the 63 reported killed between 1837 and 1840? There may be a simple mistake here, and that these totals include other species and are aggregated from a greater number of years, perhaps 1837 to the 1860s. If my hunch is right, that these raptor records from Glengarry and Glenquoich cover a greater period of years than the compilers of the book managed to translate, it would average out at a few Goshawks per year.

* * *

The site today has a modern building in a clearing framed by the straight edges of young spruce plantation, a scene greatly changed from 150 years ago. But somehow it isn't difficult to imagine Murdoch with his shotgun, and the Goshawk tumbling out of the sky. Without shooting and killing both it and the Blackcap, the keeper would have had no proof of what he had seen, and no record at all would have been noted. Why it would have responded in such a way to the sound of the first gun blast is a mystery – it is surely untypical for a raptor to approach the sound of gunfire.

A short distance further on I reach the loch, steely grey and black, with a fish farm, a goose. Happily, I also find Roddie's scoters, little dark ducks, easy to miss, just visible as the wavelets come and go. These are rare birds now, much threatened by developments at sea. Further confirmation that old Roddie knows what he is talking about.

A little further on is a popular spot for motorists to stop and take photographs. Loch Garry in its glen to the west towards Glen Quoich forms a shape rather spookily like that of Scotland itself, complete with its diagonal fault lines and coastal indentations, splitting the country roughly into three distinct blocks. There are several camera-wielding motorists, clicking away, engines still running.

I turn the handlebars east again, and it doesn't take long to get back from here on the main road down the glen to the visitor centre. I find a note from Stroma, the historian. She has unearthed some more information about Murdoch Mathieson for me.

"I'm not sure the killing of the last Goshawk is what we want to be remembered for!' she writes. "Typical of the attitude of the time, though. All the best in your research."

"I appreciate that,' I reassure her, in my later reply. 'But the way to think of it is that it was the practice of the period, as you say, and it underlines the potential of the glen to have these species again, and while of course the composition of the woodland is very different, we know that Goshawks and Red Kites, for example, can thrive in similar places. I hope the glen will one day be known for getting them back."

Other authorities think the last breeding Goshawks were killed at Macbeth's Birnam Wood near Dunkeld in Perthshire, in 1883. I have heard that these birds are still kept, stuffed and mounted, in Perth Museum. I make a note to visit there at the earliest opportunity. I'd very much like to see them.

I'm intrigued by those old keepers' lists. At face value, the volume of raptors looks unfeasible, and we have to be cautious about the identifications given. But the numbers are explained by two points: first the fact that so many keepers were employed and killing things was one of their main jobs; second are what Mick calls "source populations".

"The kill tally does not reflect local population density or productivity," he explains. "The main effect of killing is that it creates a hole in the population into which flow all potential recruits from whatever distance dispersers use – the 'sump effect'." This effect is enhanced because gamekeepers not only kill predators but both manipulate habitat and release game birds. This promotes an artificially high density of naive prey, which attracts (and holds) birds of prey. Dispersing or itinerant raptors wander until they come across good feeding opportunities. They orientate towards areas that offer the best of foraging opportunities and then settle, until they are removed. Removal keeps the sump effect working, so where the source population is large, the kill rate is determined by the rate of removal. Glengarry was no paradise, it just seemed better than elsewhere for a wandering or newly arrived raptor."

As for the apparently high number of Goshawks, Mick thinks that in fact they are low. "Goshawks wander widely, selectively home in on game areas, and are very, very easy to trap. If a Goshawk is regularly using a release pen, a box trap will often get it within a day, mostly within three days. I am surprised that more Goshawks were not killed at Glengarry – the low numbers probably reflect the bird's scarcity across Scotland at that time."

We know that much more game was capable of being reared at that time. North-western Scotland has an infertile bedrock. Professor Ian Newton has pointed me to a study by McVean & Lockie (1969). They explain how the forest cover on this land can "be utilised by the controlled removal of timber and the harvesting of wild animals such as deer. Provided this exploitation is carefully regulated, the ecosystem continues to yield a quantity of protein and wood products which can be surprisingly great in view of the poverty of the environment. Such an ecosystem keeps available nutrients in circulation and the addition of nutrients from rock weathering and biological nitrogen fixation at a high level."

Nutrients can also be added from the atmosphere, they say, "but the retention of these nutrients depends on the activities of an intact soil and vegetation complex ... Attempts to increase the rate of exploitation and utilise the resource capital directly by clear felling, extermination of wildlife, and stocking to the limit with domestic animals can be disastrous."

In the wet climate of the north-west, saturation slows down the nutrient cycle, there is more leaching or locking away of nutrients in peat. Burning results in further loss. Then there are the thousands of sheep, with crops of carcases and the nutrients they contain being removed every year. The result is that in less than two centuries of excessive land exploitation, western Scotland has been turned from a fairly productive self-sustaining ecosystem, supporting a lot of life, to the barren, degraded and impoverished landscape we find today.

***

Roger Lovegrove has given me some further insights. "Not only were there numerous keepers, but the estate at the time was enormous, straddling the whole of that area of the Highlands," he reports.

Roger has also tried to benchmark the Glengarry records against those of other estates. Some were inclined to be helpful. "I searched endlessly through other estates for gamekeepers' vermin records, but with little success.' One estate told him that they had three pantechnicons full of estate documents, the weight of which had brought down the ceilings in their castle, 'and they were not inclined to start searching through them."

"Someone with youth and energy on their side may one day look even deeper, because I am convinced there must be some valuable records somewhere", Roger adds. A fascinating project for the right person.

The lack of detail in widespread circulation about how the Goshawk came back to Britain is interesting, especially when one considers the continued blaze of very welcome publicity that surrounds the recovery of other species.

Perhaps uniquely among the increasing number of success stories of conservation reintroductions, the assisted return of the Goshawk to the UK has been shrouded in mystery, occasional resentment, secrecy, hearsay and speculation. Most books appear unclear on the extent to which the return was due to deliberate release of birds, or accidental loss of birds that then became feral, and fully wild, much as T. H. White's may have done.

"Lord Portal told us that most of the hawks and falcons trained for falconry eventually escape," wrote Eric Hosking, after his visit to see the Sussex pair. "And it was obvious that a pair of goshawks had come together."

If this could happen in Sussex, in densely populated, lightly wooded post-war southern England, then why not elsewhere? If Lord Portal was right, and there is a certain inevitability to the losing of hawks, particularly the hard-to-train-and-tame Goshawk, then did any reintroduction scheme have to be more systematic to work?

We might suppose not, but is there any way of checking? Fortunately, there is. At least one of the unofficial reintroduction schemes can be looked at more closely. Goshawk expert Steve Petty has provided me with the information on this. I had hoped to meet Steve, as part of my Highland tour, after Glengarry, but he was taken ill, surviving a heart scare while out in the hills. I am grateful to him for his help with my enquiries, and happy to report his recovery.

Steve has carried out a study into one of the more systematic attempts to reintroduce the species. This took place amid the scenic surroundings of Argyll, in north-western Scotland, between mountains and sea. Interestingly, and perhaps surprisingly, if we take at face value the numbers of Goshawks reputedly killed at nearby Glengarry a century earlier, the scheme failed.

Five male and three female Goshawks were brought from Norway, Sweden, Germany and Finland and released between December 1969 and 1973. The ones that had been taken from nests as well-developed chicks were introduced to their new Scottish home by being placed on platforms in forest trees around the Fairy Isles of Loch Sween in Knapdale. They were fed there until fledged and independent. Others had been caught alive at game-rearing stations. One of these birds was shot dead not long after release, having killed a chicken.

It would of course then be difficult to know exactly how these birds were adapting to this new environment, but one nest with young in it was found in August 1973 in a forestry plantation. The young from this nest were seen to fly on 12 August. There were sporadic sightings of Goshawks in the area in later years, but no other evidence of breeding.

Another three birds were released in the summer of 1980, and again a nest was reported, eight kilometres from the release site. This one had three well-developed young in it. After that, no trace of breeding by the birds was found, and the scheme is believed to have failed to re-establish Goshawks in this part of Scotland. Steve thinks that the most likely reason is that there's not enough here for Goshawks to eat, particularly in winter. There isn't much in the way of avian prey in these often ecologically degraded forests, and Goshawks may be forced to move on to find it. Around ten records of Goshawks per year were submitted up to 1977, most often from islands like Jura and Mull, although as usual it isn't certain how reliable these reports might be.

Mick adds a more general point about the likely fate of the introduced birds. "The failure of most of the small populations of Goshawks resulting from releases and escapes in the 1960s and 1970s, is largely because of them being trapped and removed. If a Goshawk is released into an area, it will wander widely then rapidly home in on the bit of the region that has slightly more prey, or easier prey (release pens, or managed grouse moor). That is why the species is so vulnerable. Goshawks became extinct in the first place because of this."

It is recorded that similar schemes were being carried out at the same time in Speyside, further east, and in the south of Scotland. Young Goshawks were placed in the nests of Buzzard and Sparrowhawk, in a process known as cross-fostering – the foster-parents rearing these nestlings as their own. Overtime for a male Sparrowhawk, you have to imagine.

The Goshawk was the first raptor species to be wiped out in the British Isles, and it seems fitting that it was the first to be reintroduced. There was a scheme at that time to bring back the White-tailed (or Sea) Eagle, but this would prove ultimately unsuccessful. With hindsight, putting the young eagles imported from Scandinavia on the remote and exposed outcrops of Fair Isle was probably doomed to failure.

One thing is certain, which is that there was no shortage of Goshawks to be brought from abroad. There was a scheme in which the buyer imported one bird for falconry and one for release – buy one, set one free.

More than 500 Goshawks were imported under license into the UK between 1970 and 1975, and between 1970 and 1978 as many as 25 Goshawks were finding their way into the wild here annually, whether set free, or absconding. That's 225 Goshawks in nine years.

* * *

Why was it done? Mick: "The motives were two-fold and simple, the desire to replace a lost species and to establish a harvestable resource. There were no official guidelines for re-introductions back then. Introductions, including of alien species, were commonplace.

"Releasing Goshawks was not then illegal. There was one particular group of folk – a 'club' that was introducing Polecat, Eagle Owl and Goshawk. Certainly the birds released in Northumberland were to establish a harvestable source, and the Goshawks breeding in Moray were at that time being harvested.

"Most Goshawks were probably escaped birds, but it was said that many were released by importers. One importer in Northumberland was said to have sold most of his imports, but released any bird that was very difficult to handle.

"My involvement started with locating birds breeding in the wild, but as I moved to south-west Scotland, I helped monitor the three released birds there. Robert Kenward used his experience with captives to study Goshawk behaviour. He devised an experiment to release hawks under controlled conditions (radio-tracked birds) to monitor their ranging behaviour, diet and subsequent fortunes. All published. I was present when birds were released on Speyside, and was joint author of the publication 'What happens to escaped or released Goshawks', but the work was primarily Robert's."

Since the late 1970s, Mick has been solely preoccupied with wild birds. He published a section of a symposium paper called 'Understanding the Goshawk' on the evidence that the majority, if not all, of the Goshawks in Britain then were imports, not immigrants.

The then Nature Conservancy Council, a precursor of Natural England, was involved in some of the work. Staff from there had taken part in a reintroduction of around a dozen Goshawks in a forest in Dumfries and Galloway in the early 1970s, on Forestry Commission land. It's not certain whether this particular scheme worked, if any of the birds survived. They had been imported from Scandinavia, probably Sweden and Finland, hawks trapped at game rearing stations, and imported to the UK rather than killed, as would have been their fate otherwise. They were simply given a good feed to set them on their way, and released to the wild, to take their chances in a south-west Scotland plantation woodland, in Shakespeare's words, "To prey at fortune ..."

***

It's interesting to consider what would have been done if the Goshawk hadn't come back to Britain the way it did. I am sure that conservationists would have pushed and government felt compelled to reintroduce them. We have obligations to restore lost wildlife, under biodiversity conventions signed up to by national governments. The world is watching. If we expect developing countries to preserve native forests and to co-exist with big cats and the like that eat cattle and occasionally eat people, they are entitled to ask us how we are getting on with our wildwoods and birds of prey. And they do ask.

Mick agrees: "If the Goshawk had not been reintroduced unofficially, it would have been a prime candidate for an official reintroduction. It ticks all the boxes as a candidate, the only difficulty being that the causes for its extinction still exist. That was also true for the Red Kite so I suspect it would have gone ahead anyway. One point of interest would have been the size of birds to be established – big northern birds, or smaller central European ones? I could make a compelling case for either – and both."

It's tempting to suppose that the larger, Scandinavian Goshawk would be suited to Scotland's northern forests, and less so to the landscape of the lowlands.

Friday

Ayrshire. I'm back where I spent most of my youth. Back then, even the grown-ups would have little sense of all this Gos reintroduction intrigue going on around us as the extended Jameson clan headed north (as usual) for a long (my folks were teachers – they had as much time off as we did – and even more appetite for the gypsy lifestyle) family summer holiday in a clapped-out Bedford dormobile van with a dashing red stripe along the side. There were six of us – three siblings, two cousins and me jemmied in for good measure. And it was right about this time, as the covert operation to put the Goshawks back in Scotland's forests is under way, that my own fascination with birds of prey ignited. It never entered my mind, till now, that one of those early settler Goshawks might have crossed our paths as we tootled along between campsites, beaches and youth hostels, singing a range of Scots/Irish folk songs and ditties learned by siblings and cousins at Scouts and Guides. But an abiding memory of those holidays is the Buzzard, and longing to see eagles.

I first met the gaze of the Goshawk in the Hamlyn All-Colour Paperback called *Birds of Prey*. The cover shows a pair at the nest, the male bringing a Song Thrush in as prey. The female is gaping slightly, perhaps calling, and there are two small downy young. The adults have demonic, blood-red eyes, and that severe, distinguished hawk expression. It was clear to me then that there's something thrilling about Goshawks. Newly hatched chickens know this, and so do five year-old children.

In thinking back to the Buzzard incident, I found myself wondering what the others might remember about it, if anything. Had it made as much of an impression on them? Would they even still remember it? It turns out that they do. The bird flew up from the road ahead of us, dropping a tiny Rabbit. We gathered around the bunny, while the Buzzard perched nearby on a fencepost, waiting for us to leave. It made a lasting impression on all of us. My brother Kevin, sisters Clare and Brigid, cousins Peter and Catherine Buchanan, Mum and Dad all remember it clearly, although there is some confusion over whether it was our holiday in north Scotland in 1972, or Mull in 1973.

My sister Brigid lives in Ayrshire with husband Duncan, and three children. From their base I am able to revisit the playgrounds of my youth, a meditative and in places disorienting experience. Most of those playgrounds, the post-industrial brownfields of the winters of discontent and the early years of Thatcherism, have been cleared to make way for modern housing and car parks. Identifying relic streams and stone bridges within the sprawling estates feels like archaeology, an unguided prisoner visit home. I eventually walk clear of the greatly expanded but still small town. The farmland rises to a crest. From here I can see across to the Firth of Clyde, to Arran beyond. To the east there is a distant woodland, a place that as a child I used to long to visit, that always looked so much more inviting than the dairy farmland on our doorstep. But it felt then too far, too hard to reach, even by bicycle.

I've checked for any historic records of Goshawk in my home county. *The Birds of Ayrshire* (1929) reports that in the 1842 Statistical Account of

Scotland "we find a list of birds vouched for by Mr John Jamieson, of Kilbirnie, 'a skilled and enthusiastic ornithologist'. In his list of many rarities, all given in their Latin names, we find *Falco peregrinus*, *Buteo vulgaris* and *Buteo palumbarius*. So there can be no doubt that the Peregrine Falcon and the Goshawk have not been confused, as was often the case. We can trace no more recent record."

Kilbirnie is just up the road from here – just beyond those mythical, distant woods of my youth. Perhaps I'll get there one day. It's too far to walk today. Instead, I'm going to try to find again a beech I whittled my name into on a spring day 30 years ago. The stand of trees is still here, and the old walled garden. I recall being halfway up the tree when carving, straddled on a horizontal limb, my pal Sammy whittling on a lower nearby branch. I check a few of the beeches, eventually coming to one some distance from my recollection of the precise location. It seems I've got here just in time. The massive old tree is now prone, rootball too shallow to withstand an unusual weather event, I would guess an autumn gale. And there in the tangle are our imprints. The carvings are now blurred. Sammy's initials SH are better preserved than my CJ 80, time having turned my neat curves into non-specific smudges. It feels odd to have got here just in time. No doubt the tree will soon be sawn up for timber, or tidiness.

I make enquiries with colleagues on the current status of Goshawk in Ayrshire. Have they got here? There are thought to have been some recent incursions by Gos in the south of the county. There isn't time for a site visit. Instead I go at last to the forest on the horizon. Closer inspection reveals it to be an estate, with towering trees, and pockets of wind-thrown firs. It has Buzzards now, and it looks like Goshawk country to me now too. I can't be sure if they have reached here yet.

Walking Brigid's children to school, we see a huge raptor beating sluggishly over the hills beyond the playground. It's an Osprey. On the road south I stop off in Burns country, to visit the recently opened visitor centre at the poet's cottage. I wonder if Burns knew the Goshawk.

A final note from *The Birds of Ayrshire*: 'It is interesting to note, however, that the Goshawk was observed in Kirkcudbrightshire about the middle of last century, when its nest was found on at least two occasions.' (It cites *The Ibis*, 1865.)

My final appointment in Scotland is in the forests of Dumfries and Galloway, in the south-west, towards Kirkcudbright. Leaving the gentle rolling hills and uncommon sunshine of south Ayrshire I take a back road across this vast area of largely uninhabited land. Sheep moor and forestry blocks dominate the landscape. It takes longer than I estimated to get across these uplands. It is early evening by the time I meet the next Goshawk man.

Ciril Ostroznik, like his father before him, is a forestry man, and a life-long conservationist. He is a tree expert, a bundle of energy and enthusiasm. We head straight out in his truck to visit some Gos sites. In contrast to David, Mick and Malcolm, Ciril's methods are direct. There's no picking lightly through the forest, or whispering. There isn't time. He wants to check a site which has been clear-felled over the winter, to see if the birds have nested again nearby. The forest here is lush, much mossier than on the east coast, steam rising through the sphagnum, an assault course of wind-thrown trunks and towering rootballs, puddles and day-glo grasses. I'm picking through this, looking out for traces of the birds, while Ciril is making a beeline for a particular spot. I get the feeling he knows which tree he would have chosen, if he were the bird, and sure enough, there is the nest; huge, but surprisingly easy to miss.

Ciril straps on the climbing irons and, to avoid marking the tree or knocking off any of its lower, dead growth, he scuttles up a neighbouring tree trunk, leans out over the 70 foot drop between me and him, and inspects the contents of the nest. "Three eggs!" he calls down, clearly pleased. "They're clean, as well. Could be about to lay more." A hawk calls briefly from beyond the wood. I find some prey remains, and a Gos primary feather. Not that we need them.

Like others I've met, an increasing portion of Ciril's time is taken up with looking after Ospreys, and even Honey Buzzards now too. He builds nests to encourage the birds to adopt certain areas. He has also built nests for Goshawks that have lost their first nest. Inexperienced birds sometimes use unstable bases, or old nests get too heavy for the branches that support them. Ciril's nests are usually used. Without this help the birds will sometimes try again, and build what's called a 'frustration' nest, maybe laying a replacement egg in it. This is often abandoned, as it's too late in the season to have any chance of success.

This year he knows of 20 active Goshawk nests, although he could find as many as 50 if he had the time. Some are as little as half a mile apart, although two miles is more usual, in well sited, mature forest. His father Hermann was instrumental in enabling the Gos to return here. He also found the first breeding pair in Wales, as it began its recolonisation there.

Ciril walks up several more trees, until the failing daylight makes it impossible to see into nests. He makes a maximum of three visits per season to a given nest, to minimise disturbance, and chicks are ringed when they are more than two weeks old. He finds a couple of the nests empty. That may change. The birds are nesting two or three weeks later this year, he estimates, probably because of the long winter. We find three eggs in a site that's been used for five years, although the male has changed in that time. Ciril thinks the male is often replaced. Perhaps their life expectancy is low, given their lifestyle. One nest is close to a pair of Ospreys. He thinks there is no conflict, as a rule. "I've seen a Gos fly at a perched Osprey, just to let it know who's boss," he tells me, "but the Osprey is a strong bird. Goshawks don't usually attack their nests."

He's also known a pair of Greylag Geese to take over a Goshawk nest. The Goshawks simply nested elsewhere. Goosed, presumably. It adds further fuel to my speculative pet theory that the word Goshawk may not be a corruption of goosehawk after all.

The Goshawk is doing pretty well in these extensive state-owned forests of south-west Scotland, but there are still challenges. Ciril once found a male

Gos in a crow trap, which he was able to release. He thinks the trap was illegally sited. He also knows enough game-rearers to know how they feel about the bird. "They just don't like having them around," he says. "Some think that just the presence of a Goshawk discourages the game birds from flying." I've heard this said about grouse moors too, that the risk from aerial predators keeps grouse feet planted on the ground. Will the Goshawk get out of the woods here? The signs are promising. Birds have been seen hunting in urban areas, including attacks witnessed on Rooks on school playing fields. They may even be able to move out of south-west Scotland and over the Irish Sea to reclaim Ireland.

Saturday

In the course of my tour I've been trying to apply some of what I've learned, and explore a few likely looking woodlands. In one, I find some fragments of eggshell, and realise that the rest of this large, pale egg is buried in a shallow pit, between the wood, a huge pile of logs and the public track. At first I wonder if it's a bird of prey egg, and that maybe a predator like a Pine Marten might have stolen and then buried this, for later retrieval. But it dawns on me that one of the illegal methods of targeting predators like Foxes and crows is to inject eggs with poison and set them out like this. Of course the egg may be as likely to be found by a domestic pet as a wild mammal or corvid. I realise I shouldn't have been handling the egg, as the poisons used can be extremely toxic, even to the touch. I take photographs and report it to my colleagues in Inverness.

Late in the evening I am visited by police at my guest house, and give a statement on what I've found. The interview lasts an hour or so. It gradually dawns on me that the police are either having difficulty believing that anyone could have found a buried egg in a forest, or, more likely, are keen to forestall any such disbelief further down the line, if my discovery is to lead to any further enquiries, searches or even arrests. A typical defence would be that evidence has been placed. On reflection, I can see how it might seem implausible. Not everyone likes to 'rake in the wids' after all, for traces of Goshawk.

To cut a long-ish story short, and following conversations between the police and my colleagues, the case isn't pursued. It's quite an odd place to have found a chicken's egg, but the egg fragments are quite old and, after a year of weathering, unlikely to have retained traces of any toxic substance added. Which may be just as well for me and my ungloved hands, and if nothing else it has been a useful insight into wildlife crime investigation, and what to do when something suspicious is found: don't touch, take photos, get a grid reference, report it quickly to the RSPB and police.

Chapter 7

PHANTOMS

MAY

The tour of the north has given me new-found optimism about my chances of finding the Goshawk near home. Local reports are encouraging. These have to be followed up. That Goshawk had been seen again at the gravel pit a few miles to the north. I look again at the report, which conveys the excitement of the observer, as well as the aggravation this unfamiliar interloper has been causing within the avian community: "... vid captured today can be seen here. It is pretty shaky I'm afraid!"

It hasn't escaped my notice that if the bird had continued to coast steadily southwards it would have passed over my village ... not sure where I was at the time ...

Another report comes in, from Jamie Wells: "As I arrived mid-evening, the adult Goshawk flew low over my head and pelted south east across the pit, pursued by corvids, before storming in to the rookery in the wood south, forcing one Rook off a nest, where it appeared to land, causing mass panic amongst the Rooks. It is an impressive beast, whatever its credentials!"

The bird is an adult female, presumed unattached: if she were settled with a mate and breeding, she wouldn't be out hunting at this time. She is either a dispersing wanderer (maybe her mate has gone missing in action), or perhaps an escaped captive bird. What's easier to conclude is that the local Rooks haven't had to contend with such an incendiary threat before. Generations of Rooks have not had this pressure to deal with. For a hundred years or more our lowland Rooks have inhabited a Utopian, Gos-free world. For much of that time there were no Peregrine Falcons, which take adult crows from time to time, nor Red Kites, nor Buzzards, which can take young or unhealthy birds. But none of these is the routine catcher of crows, young and old, that the Goshawk can be. The impact of returning Goshawks on Rook ecology would be fascinating to study.

The gravel pit area in question is the sort of place someone might fly a captive Goshawk, with or without permission. It could be flown at Rabbits, but would no doubt be readily distracted by the abundant wildfowl, gulls and crows. The description of the bird hunting in the evening, in that apparently clumsy, reckless way, suggests a hawk not that experienced or showing much discretion. But perhaps by now it has worked out how to catch Wood-pigeons. All speculation, of course.

Given that this is an adult female and should be incubating, she's obviously not breeding and therefore (in my view) probably an escape. It's also possible that if there's no spare male on a nearby territory, or something has become of him, a female might wander. But the likelihood is that this could be an escape.

I go to check out the nearest suitable-sized wood. A footpath runs alongside one edge. It has trees of reasonable maturity and scale, room for a Goshawk to manoeuvre, and quite a lot of ground cover. The bluebells are in full glory. There is evidence of gamekeepering, and Buzzards seem to be in residence. I'm not able to look any more closely than this, but leave satisfied that a dispersing Goshawk might at least rest up and roost here if hunting nearby.

Sunday

Two weeks have elapsed since the last report. Life has taken over. But the reports of the Goshawk at the gravel pits continue to play on my mind. The bird may well have moved on, but I need to go and check out the lie of the land there, get a better sense of where it has been.

The pit in question is part of the network of gravel and workings that flank the River Great Ouse in this part of its shallow valley. A similar pit near here drew in the Goshawk seen a few years ago by my German colleague. Migrant and passage birds of all kinds navigate through southern Britain via this long corridor. Looking at a map of the country, it's easy to see how it might serve as a gateway for birds drifting across from northern Europe.

It's another cloudless, fiercely sunny day. The gravel pit itself is wide, and shallow, and pleasingly open, not yet clogged with weed growth. Recent drought has exposed a wide shoreline and an archipelago of mud ridges for birds to rest on and probe with long bills. The cries of gulls, geese and terns provide a soothing, seaside feel. I drop the bike and sit in the long grass, with my telescope and notebook. A Skylark rises. Just one. Feels like there should be more, in such a big space. A wagtail lights on the mud and sand exposed by the drop in water-level evident around the edge of the shallow bowl. Coots have built nests opportunistically on exposed ridges of silt. A pair of Garganeys waddle onto the shore. Tufted Ducks loaf on an island. Greylag Geese graze on a raised bank. Cormorants beat heavily overhead. All very tranquil. But this is suddenly disrupted when gulls erupt at the far shoreline, catching my eye. What's bothering them? Each other, it seems. They subside again. A passing crow tilts at them provocatively. A small knot of Feral Pigeons floats in in a tight little fanned-wing formation, to drink at the water's edge. They leave together again, sticking close for comfort, genteel day-trippers to take the waters at a spa, a little insecure among all this rural exposure.

I like it here. It feels untamed, somehow. Little visited. A place to clear the head. But not of Goshawk. I wonder if it's still out there somewhere, perched in a distant tree, waiting. This is as far removed as anywhere could be from the Goshawk places I have recently inhabited. It may lack woodland cover, but the super-abundance of prey species would have a hungry hawk's head spinning. I wonder how much these locals have already got to know the Goshawk. They must all surely have seen it around, by now. I wonder at the Coots, so exposed on their nests. Probably out of reach of Foxes, which aren't keen on getting wet, and more or less able to fend off crows and other avian raiders, at least till the eggs hatch. But a Goshawk? I've read that in the past Coots and Moorhens made easy quarry for a falconer's Gos. At least in this setting they'd see it coming from a long way off, especially with a corvid entourage to sound the alarm.

Before I leave I investigate the small spinney 100 metres from the lake's edge. I assume this has the rookery attacked by the Goshawk. On the way there I find plucked feathers of gull and crow, apparently adult, in both cases. On the path into the spinney I find more feathers – Pheasant, goose, pigeon. There is a terrific commotion from a Rook nest high in the canopy above. I keep my eye on it and a parent Rook soon makes off, noisily, having just fed a brood of even noisier, well-grown chicks. They are quite insanely raucous. Discretion is clearly not a key part of their nesting strategy. If the Goshawk is still around, it hasn't cleared out all the Rooks here. There are maybe up to 10 nests, spread around. I look for more feathers, among a thick ground cover of mainly nettles.

I move to another part of the wood, where it is easier to manoeuvre, with little or no cover on the ground. There are feathers well scattered throughout – Rook and pigeon, mainly. I find the first of several plucking sites, of young Rook. I figure this is probably mostly the work of a Fox, as the feathers seem in some cases to have been bitten off. Young Rooks tumble out of the nest to make their way in the world, and most I suspect roll straight down the gullets of Foxes. But I do find one plucking patch with the head of the young Rook lying nearby, picked clean of all but beak, some feathers, and most of the skull. Could a Fox have done that? Perhaps this was Goshawk work. I photograph the Rook head, to consult with the experts.

I recall Robert Kenward's story of a Goshawk he had fitted with a radio transmitter which he found dead, killed by a Fox. Nearby, he found the Fox that had killed it, with a tell-tale puncture wound in its chest. The Goshawk had fought hard enough to administer a lethal blow with its killing claw.

I find more pigeon feathers, more crow, and even Mallard duck feathers. And dove. These were close to Rabbit burrows. The Rabbits had been excavating an old bottle dump, with bottles and broken pottery everywhere, and rusted tins, and even a bike pump, bits of iron, and a brick wall becoming visible again within this lagomorph mining town.

I cycle a few miles, through narrow strips of trees and along mayflower-festooned hedgerows. I investigate the remains of a dead animal in a field, near a track. I find the snarling face of a Fox, contorted front legs – and no hind legs or tail. Half a Fox, severed in its middle, somehow. And not lying in the place where it had been killed. Someone has dumped it here, for some reason. I can't think of many explanations for that occurring in nature.

Monday

I'm back, first thing, to look again. Blackcaps, Wrens, Chiffchaffs, Chaffinches, all sing at full volume. There is a section of this copse that is on the other side of the stream, recently dredged, too wide to leap. I find a way over, disturbing a Red-legged Partridge. The sun is already scorching on my back. I make my way into the copse, the stream creating the sort of corridor I think a Goshawk might choose as its way in too. Big birds, obvious choice. Thinking like a Goshawk? I feel only semi-literate in the 'reading Goshawk' stakes. The Goose Grass and nettles have grown high, and my knees are stung in places as I push through. I immediately find a kill, feathers strewn around and beneath the thin branches and stems of a fallen Elder. Below it lies a pigeon, stripped of flesh and head, wings and feet akimbo. Feathers definitely plucked, not bitten off. The breast plate has been picked clean. The keel bone

– the rudder – isn't just snipped, it has gone completely. I'd guess this was done in the last couple of days. A Fox might surely have picked it up to finish it off, otherwise.

I re-examine the rest of the wood, the bits I'd skirted over yesterday, and beyond. I find a pair of spectacles. A seed-hopper. I must be searching very intently, because somehow, among the detritus of the woodland floor, I spot the head of a Woodpigeon. It is lying on its own, is fresh, and has been picked hollow. Can it really have come from the carcase I found earlier, which is a good 80 metres away, on the other edge of the wood, and across the stream? A Magpie watches me, silently, but nervously, as I contemplate this question. I wade in Cow Parsley and Buttercups, through neat rows of Ash. My foot goes down a Rabbit hole.

I follow the course of the stream, in the hot sun, to enjoy the scattered hawthorns in full blossom, peaking. An egret beats past, close to me, and drops to the stream-side somewhere out of sight, but not too far away. I sit on a low bridge and try to call Mick, to describe the pigeon kills. What would have torn the head off like that? Also the juvenile Rook. There is another commotion among the gulls, a nesting colony, now quite distant, by the pit. A large raptor is up. I get the scope on it. Goshawk? No. Like a huge gull itself, it drifts. Languid. Long-winged. Osprey. It is fast becoming the bird of the year. I wouldn't have expected to see one here. It floats slowly and circles, away and over other lakes in this pit system, looking for a few seconds like it is poised to dive, then drifts south, sun on its back.

I find more fresh Woodpigeon pluckings in the corner of the rough grassy area, near a hedgerow. Definitely plucked. And just over the hedge and sand bank I find a clump of young crow feathers and bones, on top of a roll of rubber tubing, seven feet off the ground, well out of reach of Fox.

I cycle several more miles then, to look at other pits, along the river, bankside scrub bedecked in blossom, chestnut and willow and hawthorn, well served with Cuckoos calling, and in one place a Nightingale splitting my ears it is so close, feet away, above my head, but I see not a flicker in the shadows of those thorns, till it stops and I roll away.

There is a tiny Rabbit dead by the path, already punctured in the rib cage by the Magpie that flies away, and might even have been the killer. I check all the possible plucking sites I pass. Nothing more. A lone Buzzard is inspecting the copse, and a Jay is complaining about it. Neither had been in evidence when I was in there. The Jay may have been in one of the Ivy-covered trees, nesting on the quiet. Keeping its head down.

Saturday

Rain has come, at last. I am looking across at Esme Wood again, from bed, wondering at the possibilities suggested by those Goshawk reports, and the bird last seen cruising high southward, in this direction. I gather my kit to fossick in the wood. The pasture of the estate is thronging with crows galore, Rooks feeding and training fledged young. Cow Parsley and nettles high. Rabbits scattering. Foundations of an old barn? I re-find the remains of the Rook, bones now blanched. I find other things found before, like an old bird box, with a hole the size of a cricket ball drilled in its front by a woodpecker. Nearby, these are feeding clamouring youngsters. I find one spruce tree, severed halfway up, the top half now fallen, leaning against other trees, two nests with it. I half walk, half climb up to inspect the lower of the two. Empty. Old. I suspect sabotage, but closer inspection reveals a rotted trunk. This ailing fir has snapped in a gale, despite the shelter of the wood around it. It is not the only one. Many oaks have suffered the same fate.

I find a plucking post, strewn with Blackbird feathers. Among these I spot another feather. Bigger. A hawk primary. It's Sparrowhawk, I think. I discover a dead Mole, bloodied on its velvet pelt, and thrush feathers, Great Tit feathers. Woodpigeons blunder out of the canopy. I wonder, was the Mole caught by a hawk, and discarded as unpalatable? Or simply mislaid, somewhere between the impatient, sitting female and the anxious, provisioning male. There are pellets, or castings, beneath the thick fir tree that had the

Buzzard nest when I checked early in the year, although today, peering up-wards, I can't make it out. Fragments of the shell of a large egg, unmarked, pale blue-ish, catch my eye.

I exit along the edge of an Oilseed Rape field to the oasis, as I think of it, the last vestige of the 18th century Great Marsh. Its fringe of willows is sending forth fluff. I cross the stream ditch. The rain is now slipping down steadily. Robin, Yellowhammer, Skylark, Blackcap all in song. A thrush egg shell. Partridge. A tractor cab is muckspreading. I lay low. Let them pass. No sign of the Moorhen or its nest now. The surge of growth has transformed the place since I was here before.

***

A headline in the journal *British Birds* catches my eye. "Apparent nesting association of Northern Goshawks and Firecrests". It is a note by Dr. Geoff Mawson. The Firecrest is, with the much commoner Goldcrest, the tiniest bird we have in Europe. Given the scarceness of the species, and the dif-ficulty of finding them on account of their tiny size and preference for thick woodland, often coniferous, he was intrigued to find no fewer than three Firecrests singing within 20 metres of a Goshawk nest he had been keeping an eye on in the Derwent Valley, Derbyshire. It is an area with a long history of crimes against raptors. The young Goshawks vanished from the nest just a week after hatching, but Dr. Mawson continued to monitor the progress of the Firecrests.

A year later Goshawks nested about a kilometre from this site. A nest-watch scheme was set up, to once again keep an eye on the progress of the hawks, and this time four singing Firecrests were recorded within 30 metres of the latest Goshawk nest. Checks on last year's Goshawk nesting area pro-duced no singing Firecrests, this time.

Besides being, at face value, an extraordinary example of far-sightedness on the part of these tiny birds, to stick close to a large raptor that would more or less guarantee security from a range of predation threats – corvids, woodpeckers and squirrels in particular – this perhaps also illustrates how the restoration of the alpha predator Goshawk to the food web more generally might put some heat on those on which it preys and that it would be apt to exclude from the vicinity of its own offspring. In so doing it might give a little breathing space to the lower reaches of the food pyramid.

With the help of supervision, the Goshawks fledged two young out of four, after the nest collapsed in a stormy night and a nest platform was hastily built by the surveillance team. Grey Squirrels comprised 95 per cent of the prey items brought in by the male and latterly female Goshawk, with Sparrowhawk, Jay and Great Spotted Woodpecker also recorded. A year later, although a male Goshawk was present on the territory, no female joined him. There was therefore no breeding attempt, and no apparent nest site focus for the Firecrests.

Wednesday

The Peak District National Park in central England covers a wide area of moors and woodlands. Parts of it have been a black hole for birds of prey for decades, despite concerted effort by alliances of police, conservationists and landowning bodies to stop the illegal persecution. The Goshawk ought to do well here, but in one year, for example, only one young was fledged from seven nests. On another occasion a crow trap was filmed, in which a Buzzard was caught. Protected species like this are supposed to be released, but when a man arrived to check the trap he was caught on videotape as he beat the Buzzard to death with a stick. It is rare that evidence of this kind is captured, but the anecdotal evidence has been that this type of behaviour is commonplace. Improvements in tracking technology and surveillance techniques are making convictions ever more achievable.

There are pockets of resistance, where Goshawks survive. I have taken another day of leave to make the two-hour journey north-west, and hook up with another team of Raptor Study Group volunteers. Mick and John are both retired, and spend most of their time on other species, where the demand and the contracts are. They manage my expectations of their Goshawk knowledge but they are glad to be able to spend a day showing me several sites they know, to check on the progress of the birds. It's a little while since they've been to them though, so I'm warned that there may be some searching to do, and some retrieval of memory data. 'Sounds good to me,' I tell them. 'I wouldn't want this to be too easy!' I'm hoping my newly acquired field skills will be put to the test.

"You can tell we're not Goshawk men, can't you?" John mutters, drily, as our search for the nest at the first site becomes a protracted business. I'm happy to be spending added time in this woodland, though. It's unlike any of the Goshawk woodlands I've been to before. It's mostly broadleaf, with a thick under-storey of shrubs. The hawks have been breeding here for five years, and are a closely guarded secret. The nest is tough to find, even though it is close to a public path. The birds have chosen a Scots Pine. It's not easy to search the surrounding area for prey remains, but I find some bits of pigeon. The young are never ringed, as that can only happen with the owner's permission, and the owner, in this case, doesn't know about the birds.

At the second site we visit, the hawks have been smart or fortunate enough to have chosen a nature reserve, although Mick and John assure me this does not mean the birds are out of bounds for persecution. There are young in this nest, and the hawks' response to our approach is not, as has usually been the case, to call once or twice to each other and then lie low, but rather it is to approach and harangue us. Both are circling low over the canopy, giving us pelters. At one stage they land on branches a matter of metres above our heads. If we were armed and minded to we could easily have swatted them. It isn't hard to see how the bolder, feistier birds can soon be weeded out of a population. We leave quickly, to ease their stress and minimise the chances of unwanted attention being drawn to them.

The last site we visit requires two hours of searching, which turns gradually into genuine anxiety that the nest may have been interfered with. When we finally locate the nest there seems to be evidence of trampling around its base, and possible climbers' spike marks on the trunk of the tree. There are plenty of crow and pigeon prey remains scattered, but no sign of the birds, bar the odd shadow shooting across the forest floor, some distant calls and some crow complaints. We have almost given up when two large hawks sail into the forest just below the canopy: great, brown, streaky forms, chasing each other. It is two fledged young, and they appear to be unaware of us. I don't know where they've been hiding, but they are clearly alive and well. With their arrival we hear some grumblings from the adult birds, and so we beat our retreat. It has been a superb encounter with the birds, well worth the wait, and a great relief.

JUNE

Ben from *BBC Wildlife* magazine has been in touch. Apparently my essay about the Goshawk in Berlin has won me Nature Writer of the Year. The reward for this is an Earthwatch research expedition by boat to the Peruvian upper Amazon, late in the year. I share the good news – and the essay, called *Phantom* – with Rainer in Berlin. "Very nice writing – congratulations!" he replies, adding a welcome update on his study. "My research about Goshawks is going on this year, maybe due to the harsh winter, reproduction was quite low. The territory where we caught the juvenile female was not occupied this year, the old nest had collapsed in winter."

I'm especially pleased to have been able to promote the cause of the birds. Suddenly lots of people want to know about the Goshawk, especially the urban Goshawk. The BBC Natural History Unit has also been in touch. They are keen to film Goshawks and tagging/tracking studies, how the birds are adapting to the human environment. I pass them Rainer's details, although he isn't aware of any footage having been captured of the Berlin birds.

Sunday

My thoughts are turning, as they often do, to Ireland: what the history of the Goshawk might be here, and whether it might be finding its way back. Perhaps the slowly establishing population in south west Scotland might provide a source of young and roving birds to make the short Irish Sea crossing to explore opportunities there, and try to gain a toe-hold. Visiting birds from Scandinavia might also be regular, if not exactly numerous. And then there are the inevitable absconding austringers' birds.

I have made some enquiries before visiting. "Goshawks are of particular interest here, mainly due to the lack of information on them," Brendan Dunlop of the Northern Ireland Raptor Study Group told me. "We would be happy to help out if we can."

Dr Marc Ruddock, Secretary of the Group, also told me: "It is thought that there might be as many five to ten pairs. Each year members of the Group report sightings to the database so that we can try to establish where the territories are. Over recent years, there have been localised sightings from several counties."

Before meeting up with Marc and Brendan I have consulted some old texts on the history of the species here. There is fossil evidence from a place called Mount Sandel on the north coast, from 7,000 years ago. Hawk is *Seabhac* in the Irish, pronounced Shouk, and instances where this word is used in connection with woods, such as one 12th century poem, and where Sparrowhawks are also mentioned by their own and quite different Irish name. It is clear the authors are referring to *A. gentilis*. Variants on the word Shouk survive in a number of formerly wooded places, so traces of the bird remain, if not the bird itself.

Gordon D'Arcy's book *Ireland's Lost Birds* contains a mine of priceless historical records, including convictions for black market trading in Goshawks from Ireland in the 13th century, at which time the birds were prized and flown by Irish kings. Irish Goshawks were especially valued not only in Britain but elsewhere in Europe, from an early era. Ware's *Whole Works* records that "those that breed in the North of Ireland are reckoned the best in the world," while Blome in 1686 noted more specifically that the best would be found "especially in the county of Tyrone". Thomas Molyneux described the woods of Leitrim as being "full of large and excellent timber: and well stored with excellent Goshawkes," while, a century before, the O'Reyleys of County Cavan were expected to pay Queen Elizabeth I annually "one sound goshawk".

Ireland's remaining forest tracts were greatly diminished during the 18th century, and there became no hiding place for larger animals, including Wolves and Goshawks. Falconry declined as a pastime, and the value of the birds was no longer an incentive to nurture them. By the end of the 19th century industry had accounted for most of the forest that remained, and the gamekeeper's guns and snares for most of the Goshawks. The last Goshawk is thought to have bred near Derry in the early 1800s, and I note with more than a tinge of irony and pathos the symmetry of this being very close to the site where the oldest Goshawk bones were unearthed. The place of earliest record of their wild existence here is also the last.

The Goshawk was extinguished as a wild breeding bird from Ireland as early as the early 19th century, but isolated birds have been passing through the country ever since, supplemented by escaping fugitives from captivity. Hutchinson estimates nine or ten records for Ireland up to the mid-1960s. There are delightful records in the old books.

"In Ballymanus Wood, in Wicklow, I saw a young male: he flew up the wood towards me, alighting on a bare branch of an oak tree about thirty yards off. The longitudinal markings and rufous edgings of his feathers, characteristic of immaturity, were strongly defined." So wrote Mr. A. B. Brooke in spring 1870. If his identification is correct, the approach of this young bird of course suggests an escaped falconer's hawk.

There is another record from that county, scrawled by a Dr. J. R. Kinahan in his copy of *Jardine's British Ornithology*, alongside its entry for

Goshawk, which states "there seems to be no well-authenticated instance ... in Ireland". Kinahan has written: "Shot by Lord Meath's gamekeeper at Kilruddery, 1844. I have seen it fresh." He may have done, but the skin of this bird was not preserved for validation.

Another specimen – a male – is said to have been obtained in County Longford in 1853, according to Watters' *Birds of Ireland* (1853), on the authority of Mr Glennon, bird-preserver.

It's curious that there haven't been more reports of Goshawks here, over the years. But what I find even more eye-catching are the apparently authenticated records of American Goshawks here. There have been several, including two shot in February 1870, in separate places – Tipperary and Offaly. I note that this took place shortly after the American Goshawk was shot in Scotland the previous autumn. Could there have been a heavy immigration of Goshawk in North America that winter, some weather event that might support the likelihood of at least three American birds making the long ocean crossing? As far as I can tell the answer to this is no. For the record, the two early 20th century records are of an adult male 'obtained' in February 1919 at Tyrone (which does correspond with a heavy southward movement of North American Goshawks that autumn and winter) and an adult female trapped at Christmas 1935, poetically enough near the fields of Athenry, Galway – "where once we watched the small, free birds fly," I muse.

Since the mid-1960s, reports of Goshawks become much more regular, reflecting not so much any upsurge in their arrivals as a growth in interest and awareness of birds, supported by binocular and bird book ownership. These later birds didn't have to be caught or killed. There is one more record of an American bird, in November 1969, from Wexford.

***

Looking down at the province from the plane, I wonder where might a returning Goshawk survive now, in Northern Ireland. I've looked for them in forests within Belfast, where the RSPB has its HQ, and Lower Lough Erne and Castlecaldwell where the Society has nature reserves, and I looked on the map for woodland near Lough Neagh, close to where once the finest Goshawks in Europe were reputed to be acquired. I hadn't time for much of a rake.

My dad's researches into family ancestry have led him to old farms and cemeteries in Mayo and Sligo which turn out to be a stone's throw from the place where Tim White reacquainted himself with the land from which his own father had come. Tim left his cottage in the heart of the English countryside and moved to the west of Ireland for the duration of World War II. Although by this time he had given up on Goshawk ownership, he was flying and sometimes casually releasing falcons, as many like him must have been doing before and to a lesser degree since.

Tuesday

I'm at my aunt and uncle's house in the foothills of County Down. I've spent some time in the early morning at the top of the hill, among the ruins of my great grand-uncle's cottage. I looked in the empty window of the cottage to find a young Jackdaw blinking back at me from the fireplace with its one good, ice-blue eye. I left it in peace, and when I looked again a short while later from the other side, it had gone. The next time I checked, it was back in place, spookily. I could hear nestling Tree Sparrows from a cavity in the one-time kitchen wall. The parent birds were coming and going with food. I sat quietly to wait for them, to get a closer look, among briers below the Ash tree growing through the kitchen floor. A male Bullfinch stopped by, foraged within feet of me, eventually noticed me, and left.

Marc and Brendan roll by on a blazing sunny day to take me to the foothills of the Mourne Mountains, and a forest from which there have been recent and convincing reports of Goshawk. Magpie remains have been found in more than one place, and what looked like a juvenile Goshawk had been involved in some kind of tangle with two local Ravens. We would search the forest for any further signs, maybe a nest. Records of breeding in Northern Ireland are thin on the ground, which isn't to say it hasn't happened. A pair is said to have bred here in 1994. There have also been recent but unofficial reports of attempts by Goshawks to breed in the Republic of Ireland.

Brendan and Marc are experienced and knowledgeable raptor men. They know species like Peregrine and Barn Owl inside out, and have been doing work on the recently reintroduced and happily thriving Red Kite. But they are modest enough to admit that the Goshawk is not yet a species they know well, or are under the skin of. Like most people, they haven't had the birds to work on, to get to know, or the time. So I find myself in the unlikely and unexpected – but not unwelcome – role of expert, able to pass on some of the insights and tips I've picked up from the real experts in recent weeks and months.

We've driven deep into a plantation woodland, along a forestry track. The wood cloaks a fairly steep hillside, and has reached a decent stage of maturity. It looks promising, the kind of place any pioneering Gos might gravitate to. We exchange some details on what has been reported from here, where might be promising, and the sort of things to look out for, then we spread out, to search. It's hard work in the heat of the sun, and the humidity of the shade. I pick up quite a few feathers, of corvids, pigeons. Some prey remains, Fox droppings. Nothing too conclusive. We explore the likeliest looking rides, gaps where windthrow might allow openings. It's an obstacle course of a forest, pleasingly untidy, neglected in feel. Secluded. There could be goshawk here, but if there are, we don't find them today. Or any clear signs of them.

Our next destination is a wood in the lee of open mountainside. Marc has studied Peregrines up there. There is a remote mountain pass where the birds began breeding again a few years ago. It is a route through which racing pigeons are expected to pass, and of course the falcons represent a hazard to them. It reminds me of a story I once heard my dad tell. He never talks much about his family background, but I overheard him telling some guests about how his own father Alfie kept racing pigeons, where they lived not far from here in County Down. I was very young at the time, and the story may not have been intended for impressionable ears. Dad always resented race days when the birds were due to return to the pigeon loft in the back yard. He and his brother would be confined indoors, to ensure they didn't deter the birds from landing, speed being of the essence as the birds were clocked in. I think he must have been at a young and daring age, but one day he deliberately spooked the birds and caused them to scatter in all directions. For that he was thrashed, he recounted, with not much trace of remorse in his voice. In fact he seemed quite amused to recount the whole episode, which I recall thinking was curiously at odds with his more general advocacy – befitting a headmaster – of the virtues of obedience in children.

I turn this memory over in my imagination as I rake in these woods, probably planted when Dad was that schoolboy. By now I have lost track of the others, and it feels like I have the wood to myself. There has been more management in this one. There are footpaths, and the sense that there is ready access here, if anyone might want to use it. The trees have been thinned, have wide spaces between them. I haven't found any Goshawk evidence, but I do hear the alarm call of something entirely unexpected. A woodpecker … surely not? Ireland, famously, has none of these. I am able to retrace Brendan, and he confirms that in fact Ireland does now have the Great Spotted Woodpecker. It has arrived. Its expansion in range and numbers in Britain has apparently forced some of them to attempt and actually make the sea crossing, which a woodpecker would need a very strong following wind to achieve. But they are here. How they will now influence the ecology of Ireland would be fascinating to chart, for anyone with the time. Perhaps we can take it as encouragement that the Goshawk is here too. If woodpeckers can make it across the water, it should be straightforward for any self-respecting hawk.

Speaking of which … I find prey remains – not just Rabbit, but also pigeon, and corvid, close together. I summon the others to have a look, excitement mounting. This looks very promising. We spread out to search the surrounding area. The trees are a good size, stout-trunked, more than capable of supporting a Goshawk nest. Scanning around, I pick up on just such a nest, further down the slope, so that it is actually almost level with us, although high in a tree. It is a little obscured by branches between us and it, but the binoculars reveal that there is a bird on this nest, a raptor. I think we've cracked it – confirmed Goshawk breeding? I call quietly to the others, who have a look too. We're smiling and nodding to each other, but need to find a better angle on the nest. If we go closer, we lose height and sight of the bird. It's sitting tight, so we do too. We find what feels like the best vantage point, and wait. The bird moves. There is at least one quite well grown chick under it, which emerges too. Now we can see them better. They are Buzzards. "Aren't they always," I sigh.

AUGUST ~ *Sunday*

Back home, I have dug out some maps of parts of the country I might be able to investigate. In this part of lowland England, woods tend to be small, angular, but regular and usually accessible to view from afar, if not completely explorable. I try to work out where the best bits of woodland are, our most likely possible Goshawk refuges. My eye is drawn to one particular oblong of green. I cannot say exactly where this is, of course, but let's just put it that it isn't that far from my local patch. I have passed this place from time to time, walking, biking, driving. The road skirts it just a field or two away. Till now I had an impression of not much more than a belt of trees, mainly conifers – pines – at least on its road-facing edge. From a distance, most things can seem insubstantial. The truth of this for landscapes never ceases to surprise me. Curiously, the map indicates a public right of way within it – a bridleway – but this seems to be suspended in an otherwise trackless landscape. The path runs along a public road to a farm, through the tiny hamlet, then through the wood, then stops.

In a spare hour, a few days on, I am able to go there to investigate. I turn off the public road and onto the hedgeless lane that cuts through wheat fields on either side to a farm. It has a large house and a group of barns and silos, on each side of a yard. As I pull up, a man is waiting for me. He's obviously noticed my approach, and his demeanour suggests that visitors such as me are not the norm. I wind the window down as he approaches. He is eyeing me and the car carefully.

"Hi there," I greet him cheerfully. "I'm just looking for the public right of way."

He hesitates.

"Is there a public right of way here?" I prompt.

"Um, not really, no," he answers, already sounding a little impatient. "Well, there"s supposed to be a path," he offers. "But it doesn't really go anywhere, people just get lost ..."

"I looked at the maps," I tell him. "There seems to be a public path that goes round that way. Is it through the farmyard?"

"It's all a bit ridiculous, really..." he says.

"Is there anywhere I can park, here?" I ask, trying to move things along where the right of way issue is concerned.

"Not really, no," he says.

I leave him to it, and the closely guarded pathway secret. I'll return when I have more time, perhaps on foot, to explore further.

I have to call the rights of way department of the local authority, to ask about this mysterious path to nowhere, from nowhere very easy to get to. The rights of way officer explains how it has come to be isolated in this way. Without the car, to reach it requires a walk along a busy highway, with no pavement and a very narrow and uneven verge, with cars doing 60 mph within feet of the walker. You wouldn't want to stumble.

I walk briskly down towards the farm. I note a absence of signs indicating any right of way, far less its correct route. So I pick my way round the back of the outbuildings, and into what I take to be the right of way into the trees. It's a lovely wood, as these things go, trees swaying in the strong sunshine and breeze, all dancing shadows and swish. The grass of the paths is fresh, protected from the drying effects of the recent hot weather by the trees around them.

I soon reach a crossroads in the path. Ahead of me a birch wood, to my right what I take to be the path with the better chance of being the right one – I don't have the map with me. A wider path leads through the pines and larches. Scanning round, I can see large nests in the crowns of some of the trees. One in particular looks Buzzard- or Goshawk-sized. Probably Buzzard, of course. There's a pond, overhung with birches and other branches, and coated in duckweed. Spent cartridges are strewn nearby, uncovered by the recent mowing. Bracken lines the route. I make a quick assessment of this as a potential Goshawk wood. Compared to woods I've seen elsewhere, further north, in Scotland and Ireland, it looks a genuine candidate. Mature trees, wide corridors, peace and quiet, Woodpigeons and crows in good numbers. The dark bulk of nests dotted around also indicates a history of use by larger birds, crows and probably raptors after them. Perhaps there could be a Goshawk structure hidden somewhere in the far depths. Beyond my reach or access, in all probability, from the official path.

And then I find the carcase of a Carrion Crow. It is headless, and picked clean. Its breast bones are shorn clean off, its wings and tail intact. A little further on I find another one at the base of a pine, not headless this time, but another dry husk of bird, its bill apparently crushed and broken, its flesh devoured. Then a third.

"What is it you're doing?" The unexpected voice in the woods makes me jump.

It's the chap from last week, again.

"I'm exploring the wood," I tell him, again, only this time I'm a little startled and so miffed. It's tempting to suspect he'd like this right of way closed down, through non-use. I leave him to mind his business.

Woodpigeons erupt from the woodland floor up ahead. Young crows are jumping around in the lower branches of an oak. I find more cartridges, discarded plastic sacks now empty of Pheasant food. Occasional butterflies flit through patches of sunlight. There is pigeon down, and one feather drifting earthward from the canopy, an echo of Berlin. It could just be from roosting, moulting birds. A shadow crosses the path. A large bird is passing over the canopy, doubled by its shadow.

Having reached the dead end of the right of way, I turn back on myself, retracing my steps through the wood. Close to the spot where the Woodpigeons had been, not far from the pond, my eye is drawn to something lying on the low canopy of bracken fronds by the edge of the forest ride to my right. Moving closer I find that it's a freshly-shed feather, lying soft against the matt green of the ferns. It is obviously from a raptor. A large raptor. A wing covert feather, not one of the main fingers or primaries of the wing, but one of the more numerous, smaller feathers that coat the upper arm. I know it could be wishful thinking but my first impression is that this is a Goshawk feather. It bears hallmarks of some I've found in other woods, in company with the experts. But it has a trace of tea-stain brown at its rounded tip. Buzzard brown? I'll have to have it checked. I proceed in good heart. The possibility, or rather the evidence, in my hand, that there are Goshawks here may just be taking a significant turn.

Chapter 8

ROVING

AUGUST ~ *Monday*

Rain has finally come, as July turns into August. This is normally my least favourite month, this period of hiatus, silly season, with most birds lying low, in moult, often non-descript weather, of lethargy, holiday, at the worst possible time. I take some consolation now from knowing now that this is when Goshawks – males, females and offspring – may actually be most active, most likely to turn up somewhere new and interesting. Here, maybe.

Goshawk. It's all I talk about now. I'm chuffed with the feedback I've been getting on 'Phantom' – the essay on Goshawk now published in *BBC Wildlife* magazine. What's most pleasing is that it's helped to generate so much interest in what I've come to think of as the 'forgotten raptor'. Now to tell the whole story, although I can't yet guess how this tale might end.

Author and BirdLife International ambassador Graeme Gibson has even been in touch, with news of his own Goshawks. "There's been a nesting pair in our woods ninety minutes north of Toronto for almost a decade. In Spring they almost define the bush; calling loudly, swooping low over our heads if we inadvertently approach too close to the nest, they certainly dominate it. A wonderful creature."

Tuesday

I have a piece of large raptor to identify. Back home I check the feather against the illustrations in my feather book. It proves inconclusive. I send photos of it to those I think might know, with a short note:

"I found this, as you see it here, on 24 July in a coniferous woodland ride. The feather is 4 inches/10 centimetres long. I've checked the feather book, which is inconclusive. Thoughts?"

Someone I know who has kept Goshawks gives it a quick thumbs-up. "Looks good to me!"

Someone else provides further encouragement. "It looks pretty good to me," he says. "It's either Buzzard or Gos. I would put money on Gos."

"Thanks," I reply, stoked. "I also have confirmation from someone who keeps a Goshawk. So, looks like I've found local Gos – official! Couldn't have worked this out without your help ..."

"Very well done, Conor," he replies. "All you have to do now is to find a nest. Good luck!"

I'm elated. Gos have been in that wood after all. I'm unsure what to do next, if anything. Share the news with selected colleagues? Or tell no one? Is it better to keep it under wraps, or try to involve anyone locally, or the owners, with the risks that might entail?

Having mulled it over, I'm not convinced suppression will help the Goshawk to colonise a given locality. Goshawks are secretive of course but I think the wrong people would find them sooner or later. There is hardly a truly secluded place left for them to nest. A better tactic than suppression might be to more actively manage nest sites: visit the landowner, encourage pride, give them advice, let them know that others know, offer to keep an eye on things ... keep an eye on things anyway. My guess is that any early pioneers get taken out, more often by shooting interests than by thieves, but of course this is speculation.

***

In the end I am relieved of this dilemma, when I get another opinion on the feather, a strong one. Dave reckons I'd got a piece of something from a bird other than Goshawk. He is unequivocal. "I would say it's a Buzzard, with the buff edge."

"Thanks Dave," I write back. "A couple of folk reckon it's Gos – first-year female. Someone has Gos feathers from her captive bird and reckons she has a matching feather."

"If you still have the feather I would let you know for definite." he tells me. "Still looks like a Buzzard to me. Gos tend not to have a ginger tip to the feathers even in first-year birds. Pictures best taken looking straight down...

"I found two first-year female Goshawks incubating this year in the Lothians," he adds. "The season ended up being quite good. Have been ringing Gos in a few places. Still none here."

I send him a better photograph. "Pic now attached with 20 pence piece for scale."

Now he's even more certain. He would, he declares, "put a week's wages on it being Buzzard."

"Interesting," I reply. "Who gets the casting vote? Mick, if I can track him down? Bear with me. I won't prejudice him by saying who reckons what! No-one's reputation is on the line here ..."

I email Mick. True to form, he comes back with a carefully considered view. I can only marvel at the forensic expertise of the man, and Dave:

"The feather looks like a Buzzard greater covert. Put the feather against a matt black background – if it shows brown, particularly towards the tip, then it is Buzzard. Adult Goshawk is grey (male more slate-grey than female) and first-year is brown but with a paler marking – ie more marked bars. Greater coverts of first-year Goshawks here have an obvious, broad buff tip.

Goshawk in your part of the UK could be of central European origin and might be darker – I am less familiar with them. Nevertheless the ground colour criteria should hold. From a first-year bird the brighter sections and pale tips are diagnostic."

"Thanks Mick," I reply. "I am sure you are right, of course. From what you've taught me re the signs of breeding Gos that you can expect find at this time of year I am starting to wonder if they are actually present at all where I'm checking, even allowing for the constraints on where I can look. There have been various sightings and rumour, but it could all be of birds not there any more. I'll persevere."

I let Dave know: "Mick thinks Buzzard too. I never doubted you for a second. But if my Goshawk keeper friend can produce a matching Goshawk feather, I'll be in touch!"

But I know Mick and Dave won't be wrong about this.

**

I have another thought, for Mick: "If I paid, could a feather like this be DNA tested? I think you said that costs about a tenner? Not that I don't think you're right, but it would be interesting to go through that process."

"DNA should be easy and definitive," he says. "Natural Research, based near Edinburgh, are doing most DNA on Goshawk – down to individual recognition. I think they are working with a lab in north Wales. Tell them what level of ID you want (species) and ask for advice."

I get in touch with Natural Research. "We receive the feathers categorised by species and they are then sent to the lab," they tell me. "They run a genetic profile that allows us to identify an individual from that species. The species we are currently working with and therefore have individual genetic profiles for are White-tailed Eagle, Golden Eagle and Goshawk. We suspect looking at the photo it is from an owl species likely, Tawny or maybe a Buzzard."

There is a later message from them. "Emailed several of our guys and the replies were generally in favour of a Buzzard covert. If you've shown Mick I assume we can rule out Goshawk. A quick comparison with my stuffed Buzzard shows that the feather is almost certainly a greater primary covert off the left wing, but perhaps slightly the wrong dimensions – proportionately a bit narrower I think. Also Buzzard feathers appear to be less square ended than the one in the photo. So I would also be wondering about Goshawk. If mystery remains, I suggest sending photos to the national museum in Edinburgh. So although a few others also agreed Buzzard covert I guess it's not 100 per cent. Sorry we can't help further."

I have belatedly recalled that there are some stuffed birds at The Lodge, souvenirs from investigations cases into illegal trade. There's no Goshawk, but there is a juvenile Buzzard. I have a riffle through its plumage and, sure enough, find a perfect feather match – a wing covert, just as the experts have concluded, from my photograph.

Wednesday

It's been another useful exercise to go through, but the feather saga has left me a little deflated. But I also have renewed determination to thoroughly explore some other scraps of woodland close by. And so on this wet Wednesday evening I park up, jump a gate and make my way across a patch of heath, Rabbits scattering in advance of me, bolting for the brambles, Bracken and burrows. Swifts are still present with Swallows and martins, but will be off again any day now.

The tall pines beyond are widely spaced, laced with birch and mature oaks. The bracken stands thick and tall between them, a stout latticework of fresh green. Below the sycamores there is less growth, more space to manoeuvre. It is easier to scan for debris. The warm and damp evening produces everything a wandering or even a resident Goshawk might need – mature pines, widely spaced, a fire break beneath a pylon, pigeons and Rabbits by the dozen, and, best of all, the slow but steady arrival of trailing flocks of

Rooks and Jackdaws, and no doubt the odd Carrion Crow, merging together into a spectacular, clamorous roost of what I estimated must be a thousand birds. From time to time something spooks them, real or imagined, igniting a nail bomb of corvids from within the canopy.

I pick up the odd pigeon and crow feather, Rabbit skull and leg bone, with traces of matted fur. I find devoured crow, showing the tell-tale pale wing patches of juvenile feathering, below an oak. Nothing conclusive. Jays squall, woodpeckers clack, and Magpies provide a thriller soundtrack rattle of consternation. I think it is only me that they are objecting to, but who knows what else might be lurking between the boughs of these dense old trees. But if the Goshawk is out there, I can find little definite trace of it.

My spirits are raised when Mick lets me know how the Gos breeding season has gone in his study area in north-east Scotland.

"This year was eventful. I missed out on much of the Goshawk breeding season due to family illness. We had a slight population increase but very poor nest success. About a fifth failed – probably a consequence of the bad winter, though persecution and timber operations got some – but average brood size. About 75 young fledged – enough to fuel additional population increase next year. I'm now searching remaining woodland for evidence of birds – mainly dropped body feathers, but occasionally I get a new nest."

"We search in the vicinity of Pheasant release pens because this is where many Goshawks get food in September. There is one pen containing about a thousand poults, which is being used by five Buzzards and three Goshawks. Losses of poults to raptors are of little consequence – this release pens contain over 5,000 birds and, post release, he can lose up to a hundred in one night when the combine harvesters are working after dark. Not all keepers take such a pragmatic view about raptor predation. Some pens lose a few a day, some more."

I haven't much to report back. "I would search near Pheasant pens but I would need to have permission, I guess," I tell him. "And then the issue of telling keepers what I am doing would arise. Tricky."

SEPTEMBER

Migrant Honey Buzzards are reportedly passing over, in unprecedented numbers. I resolve to look out for them, to watch the sky more. Looking for Goshawks isn't usually a sky-watching game. For Honey Buzzards, I need to adjust my focus.

Word has been reaching us from the 'raptor camp' conservation volunteers in Malta that migrating Honey Buzzards are still being shot down illegally by groups of hunters. Perhaps these could even be some of our own tiny scattered population of Honey Buzzards, or maybe Sweden's much more widespread breeding birds. One was taken in for veterinary treatment, but couldn't be saved: a sad and wasteful end to an intriguing bird on an already improbable journey. Malta seems to epitomise all that is most craven in the shooting fraternity, and although the responsible wing of that alliance seems strangely reluctant to ever voice a public view on the matter, the illegal activities of Malta's hunters must make responsible hunters cringe. Some of those law-flouting hunters shoot anything that flies, including Flamingos, Swallows, and even butterflies. This is hunting as war, railing against anything that flies free. The anger and frustration of these men is palpable. I've been working for a decade now with partner organisations backing the RSPB and BirdLife Malta in the long, arduous and often frustrating struggle to make wildlife protection laws work there. I have nothing but admiration for the physical courage of my staff and volunteer colleagues in the face of threats, intimidation and sometimes actual violence on that small, and in parts blighted island.

Like the Goshawk, a Honey Buzzard can turn up anywhere. I met a man who has permission to keep beehives in a small spinney about a mile to the south of my house. I met him on the road there one day, and got talking about how the honey production is going. There have been recent headlines about the problem for bees of something called colony collapse disorder, which some studies have linked to a new generation of insecticides. He seems to have avoided that so far, but he did mention that one morning he'd gone in there and disturbed a Buzzard that was sitting on one of the hives. He wasn't aware of the existence of Honey Buzzards, which as the name suggests are expert at finding hives, which they do by watching and following bees and wasps. I can't be sure that it was a Honey Buzzard he saw, but I wouldn't be too surprised. They are reported very occasionally in the locality, and are cryptic enough to stay right under our noses all summer without anyone knowing.

Honey Buzzards have to be wary of Goshawks. They aren't well equipped to fend them off. They come in different colour morphs, and I've heard one theory that they are even evolving to be more often now the dark version or phase of their plumage. This makes them less vulnerable to Goshawk attention, as they might be mistaken for the Common Buzzard, which is better able to stand up to Goshawks. A fixed camera on a Honey Buzzard nest has captured the moment a Goshawk predated the young, and a study in Europe is showing that Honey Buzzards are also adapting their nest site choices for the avoidance of proximity to Goshawks. It may be that they will have to become a bit more tolerant of proximity to us.

I have now had some enlightening correspondence with raptor expert Rob Bijlsma in the Netherlands. He has been busy with research on Honey Buzzards, and has summarised the Dutch Gos situation. It seems that for reasons not fully understood the Goshawk hasn't penetrated cities as much as in Germany, and is now in decline after years of range expansion.

News has also come through of a migrant Osprey shot in Sussex. A reward is offered for information leading to a conviction. Malta isn't the only country with this problem. Meanwhile, the RSPB and others have been calling for landowners to bear the responsibility for wildlife crimes committed on their land, rather than those who may simply be following instructions.

* * *

Dad took ill in the summer, lost his mobility quite suddenly, and went into a gradual but steady decline. The last time I saw him he was in a local community hospital, caught between the unendurable pain of osteoporosis and the unpleasant side effects of pain-relieving medication. Mum and I were telling him I'd just been out in the forest, doing some Goshawk work. I've been helping Malcolm set up a pit-tag reading device, to try to identify the individual Goshawks at a particular site. Despite his own predicament Dad was upset to think about crimes against these birds. I was moved by that, although I felt at that moment it didn't seem so important.

OCTOBER

The clinging chill of morning mist, and a garden bedecked in cobwebs, give way to a still blue afternoon, webs once more invisible. I noticed a podgy Common Frog earlier. It had climbed onto a raft of vegetation in the middle of the old bathtub pond. It toppled clumsily over at my approach, like a couch potato off a sofa. It was either not yet quite fired-up enough by the autumn warmth on its back, or actually too perturbed by my presence to bother jumping properly. We've been face-to-face several times before, after all. It's a big frog in a small pond.

I am aiming to give the grass its winter cut, when it has dried off a bit. Meantime I look again for the frog, to see if it is back in place. No sign. As I peer, I notice another life form materialising. A flat, angular head is dangling over the rim of the tub, and a tiny tongue is flickering out, and in, out, and in. A pencil-thin Grass Snake is on the prowl.

To say it is inching forward would be to exaggerate the pace of its advance. It moves in millimetre increments, its tongue the main giveaway. The progress is agonising. I wonder is it tasting me on the air, or something else. Suddenly it strikes, and another frog, till now unseen by me, skims across the pond surface with a skittering sound. An adrenaline shot has turned it momentarily into the amphibian equivalent of a lizard in South America they call Jesus – or Tetetereche – after the sound it makes as it walks (or runs) on water.

The weeds by the bath shake as the fleeing frog plunges into cover. The snake looks confused, now blindly animated, head waving as it searches the pond for its intended victim. Slowly it relaxes, then continues on its tortuously slow prowl. The hunt is not over.

This being the garden, I am able to pull up a stool and watch for 20 minutes or more as the snake weaves and coils among the bathside stems: now I see its dark eyes and yellow collar, now I don't. It is making a relaxed circuit of the pond, with Gollum-esque poise. It pauses in a sunspot, absorbing some more rays. It slides on, now getting warmer, where the fugitive frog is concerned. There is another rustle, then a loud plop.

The frog, I think, will live to play hide and seek again, perhaps next spring. It may just have been too big – too hot – to make an autumn supper this year. Next year could be different.

NOVEMBER

It seems I've become the go-to guy for Goshawk chat. Message from a colleague, Mark Eaton. "Saw a Gos today – first I've seen in Cambridgeshire – while out at lunchtime. Picked it up in my scope, high to the east, it was cruising steadily westwards, stooped just enough to scare the hell out of the Lapwings before continuing on its way. A young male – maybe in from the continent?"

I get chatting to another colleague, Zoltan. He is originally from Hungary, and recently was working to ensure the massive road development linking western and eastern Europe is routed to avoid supposedly protected ancient forests and wetlands, including, as it happens, places that I hope to be able to visit in the near future. He's a big fella, Zoly, the type you'd struggle to relieve of a rugby ball, if he had any desire to hang on to it. His bright blue eyes peer from a friendly Slavic face, gruff, but always with a hint of humour.

He begins to talk about his days as a student at forestry college in Hungary. Since we're on the topic of forests, I ask him about Hungary's Goshawks. He tells me they have many. He has been very keen on birds from a young age there, and would always see Goshawks when he was out exploring. Cities like Budapest? Not so much, he says. There, the Sparrowhawks take over. And then he tells me something that interests me more than he might ever have imagined. "You know, I saw a Goshawk here not long ago."

It's like the Lars conversation all over again. I prise all the details out of him.

"It was about a month ago, I was driving, but my girlfriend saw it from the car. I hadn't seen it, but I turned the car around and we went back. I got the binoculars on it, from about 60 or 80 metres. It was a male, much bigger than the pigeon it was eating, but not big enough to be a female. It moved the pigeon further away from us. I didn't go closer, I didn't want to disturb it."

We pore over a map. I know the place very well. I used to travel that route, between here and Sara's old house on the edge of fenland. It is also close to where my old pal Alasdair lives, close to a network of gravel pits where we have sometimes walked.

Saturday

A chance to follow up on Zoltan's report. I park in a lay-by to avoid the rain-filled potholes. How godforsaken this place looks today, bleak edgelands of a small commuter town, its nightclub in an industrial unit, litter, winter wheat stretching to the next main road, and drab, rain-flecked skies. I strap on camera belt and binoculars as I wait to cross the busy road. A group of Woodpigeons detonates, sending a scatter of crisp leaves from the browned-off oak on the corner of a row of trees that we identified on the map. I pick my way along the edge of the vast field of winter wheat, via a narrow margin at the base of the row of trees, scanning for any traces of feathers. Many autumn winds have passed through in the past month, and leaves fallen.

There are two rows of tall, tight-packed cypresses, a tree-hedge, with a screening and slightly wider belt of deciduous trees, now mostly bare. Not a woodland so much as the disembodied edge of one, surrounding another huge arable field behind, following almost all of the perimeter. Peering through the other side to get the lie of the land, I can see a farm about half a mile away. It's unrelievedly bleak, on a day like this. Thank goodness for the possibility of a Goshawk.

The under-storey of bramble has withered back. It is quite easy to pick a route through this belt, looking for any traces of prey remains. A Goshawk feather would be the best form of evidence that the site is in regular use, if not a larger prey item's breast plate. I find pigeon feathers strewn around, several different birds plucked here, and spent shotgun cartridges of many colours, some faded from red to pink. There are the remains of a Jay, about a month old.

A Magpie starts up a protest, rattling throatily, but staying well clear. But protesting at me? Or at some unseen danger? Did I flush a hawk? If a Goshawk were resting up somewhere in this tree-hedge, it could easily slip unseen out of the other side. I wouldn't have a hope of seeing it. A little further on a raptor appears over the tree-tops, and moves over the hedge. I scramble to peel back the wall of Cypress branches to look beyond, and there, right overhead, cruises a Buzzard, occasionally beating its wings calmly, and I am sure looking straight at me, this odd face a beacon to its keen eyes from within this long, monotonous wall of green. A Gos would not have drifted over like this, if I flushed it. It would scarper.

I sit for a while, trying to get inside the mind of a passage Goshawk. Sitting still, watching and waiting, this is the tactic of choice for the mature Gos, an experienced hunter. Low input, almost loafing, with half an eye on the surrounding terrain. If a prey opportunity arises, close to hand, accept the gift. Let the prey come to you. Keep an eye on the field, from a concealed spot that still commands a view. The thick foliage of this hedge and the screening belt of deciduous trees might be almost ideal. Wait for the prey to gather, to venture close, before launching a sprint, an attack or chase flight. A large hawk can sustain this for about 300 metres – getting into middle-distance territory. The Goshawk can be a tenacious chaser. Sparrowhawks will often give up, if their surprise attack is foiled, especially in the open. The Goshawk may bank and chase again. It combines power with manoeuvrability, muscularity with stamina. Birds like Woodpigeons in flocks will tend to see the hawk coming soon after the attack is launched. That is why they are in flocks. More pairs of eyes on the hedge.

If nothing happens for a while, the hungry *Accipiter* will move to another rest stop and vantage point. Every movement brings the possibility of betrayal, by spying eyes. So there is risk as well as cost in moving, weighed against the risk of staying put. And with each failure, the work will get harder, the effort greater. The hunter will become less patient, and be tempted to go for less promising opportunities. It may then resort to hunting on the wing, to visiting those built-up areas with their alarming, humanoid forms.

I wouldn't have looked at this place twice, in passing, but on closer inspection it feels like a reasonable spot for a wintering Goshawk to spend a bit of time, in the context of the leafless, exposed flatlands beyond. A bird from north Europe might find these conifers somehow homely. Not native to Britain, but normal enough for continental hawks, a sheltered place to rest up, from which to keep an eye on the crop beyond, and the parties of pigeons that build up on the emerging shoots, to graze them. Definitely a base from which to launch ambushes. Zoltan must have been passing after just such a moment. There's always the chance he was mistaken. But if a Hungarian, former forest-dwelling ornithologist can't be regarded him as a reliable source, then what hope for us rookies?

'You know it is there, because you don't see it.' Rainer's words play out in my head. Perhaps there is a Goshawk still around here, somewhere. More likely it moved on soon after Zoltan saw it, successful for a few days, while it retained the element of surprise.

I get in touch with a renowned and respected former colleague, Mike

Everett, now retired, who knows this part of the world as well as anyone. He is also one of the most knowledgeable birders and probably the best sky watcher I've ever come across – acute eyesight up to retirement age, but also just the careful habit of bothering to keep looking up there all the time. He taught me the importance of watching the sky, to keep looking. His eyesight was remarkable, but so was his acuity for seeing the dot, and not taking his eye off it.

"Goshawk is of course possible," he replies, "and they are reported from time to time in the county. I've never seen one myself. There were some local records years ago, and I believe one of these was from my patch – but I suspect misidentification. I'll keep my eyes peeled, nonetheless."

***

Sunday

A check in Esme Wood is overdue. I've got renewed optimism from Zoltan's Goshawk. What is it with these pairs of foreign eyes? I have to get out at least once a day or I get cabin fever. Starlings cluster like unharvested fruit on the tallest trees of the wood edge. Ten Collared Doves sit on a lower tree by the thatched cottage, all pointing in one direction, poised, as though under starter's orders. They shoot off, with other birds in tow. A Sparrowhawk scoots low in front of me, past two men chatting outside the cottages. I don't think they saw the hawk for what it is. It disappears along the woodland path I am about to follow. Definitely not a Goshawk.

An oak has snapped at head height, revealing rot in its midst, and is lying among the irises of the pond. Someone had already been scavenging some of its branches. The cattle are bellowing like stags, suckling calves in the bitter rain. Sodden leaves are pasted to the path and woodland floor. But the walk is refreshing, energising. A Jay squawks. Small birds flit through the canopy, Treecreepers among them. Quad bikers have been using the permissive path as a rallying track. I sense that any hope of Goshawk settling here has gone again.

Monday

St Andrew's Day. A stabbing north wind brings tears to my eyes and vigour to Monday morning thoughts as I walk to the bus. I let tears roll down my nose, and feel a bit better. I've heard it's good to let these things out.

DECEMBER

Daylight has revealed the first proper frost. The half-light of winter dawn gives way to a sparkling morning. Sara left early and had to scrape the old car, inside and out, before heading off to Bedford in a plume of exhaust on the chill air. I set off in a mock sheepskin jacket and a Biggles-style furry hat, the kind that fastens below the chin. I do so conscious that I might be making my head some kind of decoy for passing raptors. There have been reports coming in of Buzzards attacking joggers. Buzzards?

If true, my theory is that they are escaped or starving birds. They might confuse a mop of flowing hair heading in the opposite direction for a prey item. I don't think the public need to be alarmed or keeping their children indoors, any more than they already do.

Winter has revealed where the Goldfinches wove their summer nests in the flimsy twigs in tree canopies. Woodpigeon and Collared Dove nests somehow resist autumn gales, perhaps too flimsy to catch the wind. If the durability of the Rook nests is impressive, this is even more so. I recall watching the Woodpigeons gathering materials in August. I could later see their eggs through the base of the nest.

***

Once upon a time there were occasional reports of hawks apparently drowning prey on purpose. The standard ornithological response to this was that the incident must have happened by accident – the hawk just happened to catch a bird in or near shallow water. Perhaps, it was conceded, if this happened once, by accident, then the hawk might learn it was effective, and do it again, on purpose. Perhaps they might even see each other doing it, and learn that way. A bit like the fashion that developed among Blue Tits, for pecking open milk bottle tops on doorsteps to get at the cream. A skill that may now have been lost, even if there were still silver milk bottle tops on doorsteps for Blue Tits to investigate. But I think quite a few people were resistant to the notion that hawks would be this smart.

That was before the era of widespread visual recording of events in nature. Now, it is hard to know whether the 'drowning' behaviour of hawks is increasing, or there are just more people reporting such things, not just by writing to or calling conservation organisations, but by posting evidence on the Internet. There is one particular example of a Sparrowhawk using water to subdue prey that clinches it for me beyond any reasonable doubt that the behaviour is deliberate, whether learned by accident or witnessed before, or not. Interesting because I have known quite a few bird enthusiasts (one very real expert has told me that birds are merely 'feathered lizards') who would not even accept that it could happen other than by accident.

The amateur sequence was shot in Germany, and can be seen on Youtube. It shows a Sparrowhawk foreshortening a death struggle with a Magpie by dragging its opponent several metres to a garden pond, and plunging it unceremoniously under the water. Learned behaviour, instinctive, or reasoned? Probably the first, is the received wisdom, though my mind is open. Maybe the only way to ever nail this would be through controlled experiments with captive birds, whose life histories would be known – ie had they ever had exposure to this before, or could they work it out? The trouble is that would require the provision of live prey, so it would not be ethically acceptable.

I'd be surprised if this young Sparrowhawk, in its short life, has seen another Sparrowhawk use water to drown prey. They are solitary birds. The chances of one witnessing another doing this at best occasional stunt are extremely slim. So are we to believe that this young Sparrowhawk did this once by accident, then remembered, while toe to toe with a shrieking Magpie?

Are we right to assume that birds such as these only make apparently clever links by accident/through observation of others, rather than through an intelligent 'working out' of the behaviour. I've come across some notes on a Goshawk nest in the Sierra Nevada mountains of the south-western USA. Here, a nestling Goshawk – a jumper, as they're known – had fallen out of the nest tree. It promptly walked to a nearby stream and had a long drink and a soak. Goshawks are known to be fond of bathing. Austringers call this soaking in water 'bowsing', from which we derive the word boozing. The young bird then continued on its walk, making its way painstakingly uphill, until it reached an elevated point and was able to launch itself back in the direction of the nest tree. The observer of this was left with a distinct impression of in-born intelligence in the bird.

I recall a visit to the Vulture Conservation Foundation based in Mallorca (no Goshawks there). They have been putting lost vultures back into the mountains of several European countries. I found among the handful of captive birds they have there for rehabilitation and breeding a juvenile Egyptian Vulture, a rare species in Europe. This one was rescued in southern Spain as it had not matured sufficiently to make the short migration to Africa with its parents. This species is known for its use of stones to break large eggs, an unusual example of tool use in birds. I asked the staff there if they had tried their bird with an egg yet. This would be an interesting experiment, in the absence of any parental guidance. The answer was no.

Perhaps the true intelligence of birds will never be fully knowable. Goshawks, meanwhile, with all the power and hardware at their disposal, would rarely have the need to drown anything. Which may be just as well because one falconer I've read, from their experience of training captive Gos, used to wonder if they even had a brain.

A vivid sequence of photographs has appeared on the Internet showing a Goshawk attacking a Grey Heron – a formidable adversary. The incident took place in Leipzig, eastern Germany, in late August. I seek Mick's views on it: "The heron is a juvenile so could have been naive or hungry or otherwise disadvantaged – nevertheless the hawk needed skill. I see it held onto the head/neck – absolutely necessary for herons. I have handled lots of juvenile herons and if you don't get the head, first grab, they easily pierce the skin with a single jab of the beak. The hawk has blood on its cere, but I couldn't work out whether this was from damage to hawk or heron.

"Not much progress here," he adds, by way of an update on his Goshawks. "Mild weather over the last ten days has lead to some birds carrying odd sticks and I heard a bird begging about a week ago – but no food being delivered to nest sites yet"

I also seek the views of Rainer, not least because this may be a new German city being colonised by Gos. "The Goshawk is a juvenile female, and the bird had dispersed from its parents' territory only maybe since 15 days," he says. "These birds are inexperienced and prone to take great risks. Second, and a little bit astounding considering the date, the heron is a juvenile as well and most probably was not even fully fledged. Probably the bird was taken from its parents' nest. Goshawks are well known for taking juveniles of large birds, in Germany also of e.g. Osprey, Red Kites or Lesser Spotted Eagles.

"I know of no resident Goshawk population in Leipzig, and a juvenile on dispersal may come from many kilometres away. But Leipzig has some very nice tracts of forests within the city and of course Goshawks may reproduce there." It's fascinating that this hawk has been out of the nest for just two weeks, and already it knows how to tackle even a young heron – otherwise it would be likely to suffer serious injury.

"Two weeks from the territory, not from the nest," Rainer stresses. "Possibly, the heron has indeed tried to stab the Goshawk, but the latter acts extremely fast, if something is approaching him, and it may have gripped the Heron's head then. And Goshawks do not release what they have grabbed once with their claws. Nevertheless, the Goshawk has taken great risks."

JANUARY

The Woodpigeon shooting season is still in full cry, with shotgun reports from local woods and spinneys. The season runs until spring. This is the time of year when any pioneering Goshawks might try to establish territory, and begin nest building. I wonder if the activities of pigeon shooters, and the sound of regular gunfire, would be enough to put off any prospecting birds.

"I don't think anyone has studied this," Mick tells me. 'From studies on other species (mainly fish-eating birds at fish ponds, and corvids in sown fields) we might expect that at places where there is lots of food and no alternative nest sites, Goshawks would rapidly get so accustomed to regular shooting that they would habituate. But if they were purposefully shot at, they might not habituate. Of critical importance here is 'choice'. Goshawks can commute big distances – usually less than five kilometres, but sometimes up to 10 kilometres – so given a choice between settling in a well-shot wood or a less-shot wood, I suspect they would opt for a quiet life.'

My friend Richard Porter has been surveying raptors in north-east Norfolk and has recorded 200 or 300 Buzzards. 'Any Goshawks?' I ask him. "Not a single one," he replies, although it was clear he hadn't expected there to be any. I have a look in the Norfolk Bird (and Mammal) Report. "In spring noted displaying at four sites ... but no proven breeding records this year."

I've written several articles about the Goshawk which are published now, and it's generated some useful correspondence. A local government ecologist gets in touch. He tells me he's seen Goshawks displaying at a site in Norfolk. I sense he is a reliable witness. The fact that he's an ecologist helps. I take the details, with a grid reference, and tell him I'll take a look the next time I'm heading up there for a weekend.

The Norfolk town Sara and I stay in is well served with antique bookshops. I find a book by a man called Lubbock – *Observations on the Fauna of Norfolk* – from 1879, a time when he still knew the bird as *Falco palumbarius*, and it had ceased to reside in the county.

"In the adult plumage it is so rare that Mr Stevenson believes the adult male killed at Colton in 1841, is the only example in mature plumage known with certainty to have been killed in Norfolk."

Lubbock goes on to list the other records of the species, all young birds, reflecting the short life expectancy of the birds as well perhaps as the tendency of the inexperienced Gos to wander into trouble.

In another dusty volume, written at the turn of the 20th century, a W. A. Dutt is unconvinced the Goshawk was ever resident in eastern England. "Whether the goshawk ever bred is uncertain," he says. He then cites a letter from September 1472 in which a John Paston makes a request from his elder brother: "I axe no more gods [sic] of you for all the servyse that I shall do yow whyll the world standyth, but a gosshawke."

A hawk seems to have been duly despatched to him, but the journey from London has taken its toll. Two months later Paston writes again: "She hath ben so brooseid with carriage of fewle that she is as good as lame in bothe hyr leggys. Wherefor all syche folk as have seen hyr avyse me to cast her in to some wood ... cast her in Thorp wood and a tarsell with hyr ..."

I have a look for Thorp Wood on the modern day map, and even go there for a cursory recce. There are caravans here now, secreted about the forest. I wonder if any Goshawks arriving off the North Sea today might be tempted to head for this spot.

Sunday

While browsing the vast second-hand book barn at Blickling Hall, a magnificent National Trust property not far from where we are holed up, I pick up a more recent title with the intriguing name *Guns and Goshawks*. The tale builds to a climactic chapter in which the author flies his hawk at wild geese. The initial lack of enthusiasm of the hawk for this project makes me wonder – and not for the first time – whether the word goshawk really does stem from the Anglo-Saxon for goose. Hawk, meanwhile comes from *hafoc*.

Norfolk has always been popular with naturalists, reflected in the large number of books inspired by its flora and fauna. Arthur H. Paterson, another author writing in late Victorian period, lists several quite different Norfolk records from the past century, including two from ships offshore that brought the birds (whether dead or alive is unclear) into port.

There are some more recent records of passage Goshawks coming in off the North Sea. On the North Sea Bird Club website, which compiles records of birds seen on and around drilling platforms, I have come across this record of an extraordinary incident on an oil rig, from a decade ago, in mid-September 2002:

"Yesterday at 1600hrs, Phil called the Control Room saying he could hear screeching noises coming from the top of the fab crew container on main deck. I raced up to the metering skid and looked down onto the container. There was a Goshawk with a Kestrel in its claws and a second Kestrel on top of the Goshawk. I rushed down to the control room and shouted for some assistance on the main deck. On reaching the main deck I climbed up on the scrap skip. I could see the Goshawk, who was actually plucking feathers from the Kestrel. The hawk then literally jumped down to the main deck with the Kestrel. Barry then tried to pull the Kestrel away with a broom, but the Goshawk then started to fly along the west side of main deck with its prey. Barry threw some pears at it, at this stage it dropped the Kestrel which surprisingly was still alive.

"We made up a box, put the Kestrel in and placed it in the workshop (time 1615). At 1900 Tony and I went down to the workshop to try and feed it with some fresh meat. After a few minutes it started to take meat, plus, at times, my finger.

"At 1930 Tony and I left the workshop on the main deck to return to the control room. We were just passing the old regen area when a voice called out, 'help give me a hand'. Dave had heard some screeching when standing outside the south smoke room door. He then raced down and found the Goshawk had caught the other Kestrel. He in turn had pinned the Goshawk to the grating with a broom, but when we arrived the Goshawk flew away leaving the other Kestrel. We then placed this Kestrel in another box. This one would not eat any meat. Tony and I then cut up some meat and placed in the boxes. We then left the workshop for the night.

The next day: "We are still trying to get a good photo of the Goshawk, which is still on the platform. I will take the Kestrels in with me on Wednesday's flight and release them at Cantley, near the woods and marshlands. Phil, Tiny, Gerry, Dave, Barry and Shaun all verified that it was a Goshawk that had attacked. At times we were only three feet from it."

There is little doubt that the behaviour sounds like Goshawk, and as many as eight observers were involved, some of whom evidently got within touching distance of the raptor-on-raptor-on-raptor struggle. The curious thing about all this is that the accompanying photographs show a Sparrowhawk, not a Gos.

I have had a chance to follow up on the tip-off from the ecologist. There is a small nature reserve here, which at least means easy access for a small fee. It's a great little fragment of wood and fen. Can it really be big enough for Goshawks to attempt breeding, though? Halfway round its well marked paths and boardwalks I come up against a path that is closed. This section of the reserve is temporarily out of bounds to visitors, with an apologetic notice. "Nesting birds of prey", it says, but not which species. I am putting two and two together here, and getting Gos.

Back at the visitor centre I get chatting to the staff. I expect they are sworn to secrecy, and perhaps it would be unfair of me to ask, even with my 'card'. A secret is a secret, and I shouldn't test their will. But there's no need. The identity of the raptors in question is revealed as we chat. They are Marsh Harriers.

The trail feels cold again.

Chapter 9

ENGLAND HAVE MY BONES

JANUARY

The snows have all but melted, bar the residues of snowmen and shov-elled heaps. Walking down the road to meet a colleague for lift-share to work, I pause to have a closer look at sparrows clamouring round a feeder in the garden of one of the village farmhouses. A car draws up behind me.

"Conor," calls the driver, familiar, although I don't recognise him. "I think I saw an Osprey yesterday ..."

I have become accustomed to and fond of conversations like this locally. I'm known as the bird guy, I think in part because of my regular column in the village newsletter.

"Not a Red Kite?" I venture. There had been reports of one over The Lodge the previous morning.

"No," he insists. "This was pure white, and really big."

"Where was it?"

He describes the place in detail. It was about 15 miles south of here. I listen carefully.

"Pure white on the front, with a dark head. It was perched near the road." I suggest a Buzzard, which can sometimes be really pale. Or a Rough-legged Buzzard, which sometimes visit in winter from the continent, and are a bit bigger than their commoner relative. He shakes his head. He's been through the bird book and ruled these out. But there is a basic problem with his Osprey theory.

"The thing about Ospreys," I tell him, "is they migrate. They should be in Africa now." I think he is aware of this. But from his description, I have no better ideas. I tell him I have to head on, to get my lift. As I turn to walk on he passes on one final report.

"I also saw a Goshawk," he calls.

I stop again. "Where?"

"In my garden – I live opposite the farm."

"Do you ever see Sparrowhawks?"

"Yes, all the time. But this was *huge*. I saw it in the book the other day when I was looking up the Osprey, and I said to my wife, 'that's what I saw that time, in the garden'".

"When was it you saw this?"

He has a think. "About three years ago."

"Do you remember what time of year?"

He has another think. "It was early spring."

"And what was it doing?"

"It was eating a Collared Dove. I tried to get a photo but it was dusk. There wasn't enough light. I could tell its size from comparison with the trunk of the tree behind it. It was a big bird."

Three years ago ... that's around the time when I was first finding prey remains in the wood, just beyond his garden, getting the odd glimpse. Perhaps there really had been a Goshawk around then.

Friday

It's dusk as I'm leaving work. I notice a crow-sized bird drifting across just above treetop level, batting its wings stiffly as it goes. I crane to see it again after it disappears behind the trees. Displaying Goshawk? I wonder if I may be starting to see things. And if anyone else noticed it.

I've been sent a clip of rare and quite recent footage from Poland. A large cow lies dead in the snow. A Goshawk crouches over it, wings mantled, as though it has killed this half-ton beast itself. A Buzzard walks into shot, on the ground behind it, while the Gos – I think from its size relative to the Buzzard it must be a male – ignores the intruder, so intent is it on separating some slivers of flesh from this huge and presumably quite tough – not to mention frozen – carcase. There's a bit of a skirmish – nothing lethal, just tetchy: like siblings at a dinner table waiting for someone to dish-up. A Raven joins them. I send this on to a few people, for their thoughts.

I know Goshawks can be fierce, and have even been used by falconers in India to hunt gazelles, with the help of dogs, but cattle are beyond their means. I like the way the Goshawk is so absorbed in eating that the Raven is practically tapping it on the shoulder. Single-minded. When hungry Gos are eating, I think all else is blanked out. I don't know why filming ends here, but I assume the Buzzard and Raven just waited their turn in the end.

The wider relationship between Goshawk and Buzzard is fascinating. For most of the year they aren't normally competing, as the Buzzard is an eater of voles, and worms, and carrion. I think an uneasy truce prevails, the Buzzard probably aware that the Gos cannot be trusted, but it's okay till you turn your back ... I have found a report of a German study by Oliver Kruger in which he looked at interactions between the two species, where both are present in good numbers. He estimates that up to a third of adult and young Buzzard mortality was down to predation by Goshawks. Earlier studies have shown clear dominance by Goshawks where there is competition for nest sites. One of the benefits for Buzzards of having Goshawks around is that there is less harassment from crows, which can also pose a threat to unattended eggs and chicks.

There's been a lot of debate about the status of Eagle Owls in the UK. Like the Goshawk, Eagle Owls are often kept in captivity and like the Goshawk, they often escape or may be deliberately released. Pairs have established themselves in widely scattered locations, and made their presence felt. They can be quite intimidating around their nests, and there are reports of dogs being attacked. This owl isn't classed as a native species, although it seems to have been present here up to around 9,000 years ago. Why it may not have survived here for longer is not entirely clear. Eagle Owls aren't great at long sea crossings and so it seems unlikely that they are now colonising under their own steam. After a period of careful consideration of the issues around Eagle Owls, the government has announced that it doesn't intend to take action against the birds that are making themselves at home out there.

The potential colonisation of the UK by Eagle Owls may have implications for Goshawks and other birds of prey. Studies in Germany have shown that the return of 'superpredator' Eagle Owls there has impacts on 'mesopredators' like Goshawks. Numbers of breeding hawks decline, until an equilibrium is reached. In short, the Eagle Owl is king, head of the pecking order, and day-flying raptors and other owls need to give them a wide berth. In the same way that Goshawks can push other nesting raptors out of their territories, so Eagle Owls take over some of the best sites. If cliff and crag nest sites are in short supply, they take over raptor nests in trees, for example.

All things being equal, it might all add up to a more functional ecosystem. Hunting mainly at dusk and dawn, Eagle Owls may take roosting birds of all kinds, and they are particularly partial to Rabbits, which weren't here until humanity introduced them.

We know that raptors like Goshawks have a strong dislike of Eagle Owls, and one of these owls – real, stuffed or plastic – placed in a conspicuous place in daylight hours, acts like a magnet for passing birds of prey. There's an interesting account of an occasion when a stuffed Eagle Owl was placed in a forest glade, intended to bring crows in for someone to shoot. The observer reported how a Goshawk was in turn attracted by the crows, and able to hunt and catch them while they were distracted by the owl. The hawk then turned its attention to the owl and began to mob it. Emboldened by the owl's lack of defensive responses, the Gos soon tore the inanimate decoy to pieces.

FEBRUARY

The snows are now gone completely, leaving damage in their wake. I trim off the frost-bitten, ragged ends of a surprising number of shrubs and plants burned by the freezing conditions. Cistus, Rosemary and Olive have been killed. But so to have even supposedly tough, robust, native plants like Yew. I am absorbing sun on my face, and hear a raptor cry. I look up, adjusting focus and aperture from two feet to 500. My vision is near perfect, according to a recent eye test, but this focusing takes longer than it used to. I raise my hand to block the sun, catching warmth in my palm. Two birds of prey are drifting high against the watery blue. Buzzards, as usual. A third bird is there. A hawk. Small by comparison, wheeling tight alongside, flickering white in the sun, close then apart, a Spitfire to their Lancasters. Sparrow-hawk. I follow its progress, knowing that if there is a Goshawk around it might be lured in for a joust. It's my best chance of finding out if any of these reported birds might be sticking around for a nesting attempt. Or if they exist at all.

The knowledge that it's the time of year to be looking for Goshawks gnaws at me all my waking hours, especially when the sun is shining. If it is, then any that are present out there ought to be spending some time in the sky, showing off, and visible from some way off. The trouble is that there are so many other demands on waking hours, and it isn't sunny that often, and only weekends are free. I can feel the time ebbing away.

Sunday

I watch from bed on bright dawns, hoping the Buzzards might attract something else in for a look round. I see another Sparrowhawk up there with them, over Esme. The Rooks scatter. Clive the farmer reported a Red Kite circling over a Fox carcase near the wood. A jet out of London to the south scorches a double groove across the sky, heading west. From time to time I spot a large shape and a momentary pang quickly subsides as a microlite hoves into view. The Buzzards are getting used to these flying people.

With warming afternoons I can lie back on the grass and watch the Buzzards wheel. They hold conventions, drawing others in from miles around, sort out their land rights, part again after some wailing and posturing. There is a raptor vision sphere that envelops the Earth, or the terrestrial parts of the Earth at least. They see so well, and so far. They travel so quickly, and most times so effortlessly. Their sense of place and distance surely dwarfs our own, parochial outlooks.

Goshawks share this space, and soar when they have a mind to. But they are not designed for this low-input idling, the way that Buzzards are. They aren't recreational soarers, as a rule. They can do it when they have to, for the purpose of advertising, defending, gaining height before plunging into a hunt, or moving longer distances outside the breeding season, perhaps. But they are carrying a lot of muscle bulk and hardware. They are built for sprints, not marathons. I follow the Buzzards until they drift out of sight into the sun, ebbing to nought way to the east. A jet roars like wind on a distant planet. I resume duties, filling the bird feeder and washing the pigeon mess out of the birdbath, before refilling with fresh water from the barrel. I note with a glow the Cowslips starting to push through. The continental high pressure is persisting. A succession of clear mornings tailor-made for displaying Gos comes and goes.

Sunday

I've grabbed a rare chance to visit the wood with the rumoured Goshawks of two years ago, where I found the feather. This time I will find a vantage point nearby to watch it, for any hawks that might be tempted up, to display. It isn't the ideal day – mist hanging, clinging damp, uninspiring, birch trees prone like bones, Fieldfares moving across the fields, with Goldfinches. Redpolls forage in the birches still standing. Crows and Coal Tits call. A Redwing lies dead in the flattened bracken, saplings sweat in plastic tubes. No Gos.

Monday

I visit the library at lunchtime. Chaffinches are pecking at the window behind me. Librarian Ian Dawson tells me they've done this every year for the three or more decades he's worked here. Same window. I find another RSPB institution, James Cadbury, browsing the shelves. I show him a Goshawk feature I've written for a journal on the shelves. He tells me that they are nesting at one of our reserves in Wales, just a hundred yards from the site manager's house. This chap apparently didn't even know about the nest until he saw the tail of the sitting bird poking out of the side of it. James has also been to Moscow, where he says Goshawks regularly eat lunch on park benches. "My wife is very keen to see one," he says.

I've been getting some feedback from the features I've written about Gos, including this from my American friend Clark, who rents a castle in Italy which he runs as an arts venue. For some years I've been nurturing his embryonic interest in *ucelli* – birds.

"Conor! Damn you, I'm trying to work on something else, and you send another beautifully written article about a bird I've never even heard of, let alone care about. Now even more of a distraction, I have to look up how to SAY it ... GOOSE-hawk? GOOS-hawk? GAWS-hawk? I wonder because there seems to be an Atlantic divide between GOOSEberries and GOOSberries. I prefer GOOSEberries as they don't sound so, well, gooey.

"It's time to gather these goodies in a book, my friend. So rarely does one find scientists willing and able to share as well as you do. And I'll be damned if a couple years ago I didn't see one in Berlin, too. My Berlin friend and I thought it had escaped from the zoo!"

"Clark, you are too kind," I say. "P.S. it's GOSS Hawk. I have my own theory that it might come from Gods (gosh!) Hawk – hawk of the gods ..."

We're visiting Aragon in north-east Spain, ostensibly to see vultures and Cranes for a feature on ecotourism in the region that Sara has been commissioned to write for *The Sunday Telegraph*. We are treated to the spectacle of a thousand Griffon Vultures gathering in a magnificent mountain location, in the morning sun, first speckling the entire sky like fruit bats, then wheeling low around the canyon in front of us, then landing to form a giant herd, surging forward like rioters when a man they know well opens a gate and empties a barrow-full of animal waste into their seething midst.

We are staying in a rural retreat where there is time to explore the pine forests of a river valley so remote there is no evidence of what century we're in. Wind whispers in the trees, around a riverbed of sculpted limestone, smoothed banks and boulders, ice-blue water. The dry brush teems with migrant thrushes, and is strewn with snail shells. Serins jangle from the pinnacles of pines. As the air warms and rises, soon after the first hawks are up, first one in a long display stoop, then a pair. Sparrowhawks again. I move into the trees from the riverbed, and near a ruined church I see two much larger hawks approaching, gliding low over the canopy. They see me and veer away. They were in view for just a few seconds. They were noticeably pale with darker tails, less dark wings. I am letting myself believe they might be Goshawks. Or were they Booted Eagles? Two crows arrive promptly on the scene, passing high, complaining vigorously, following the route taken by the unidentified raptors.

I later see another hawk scooting high over the valley, wings folded, ragged edged, aimed at distant crags. A Raven passes high. Griffon Vultures wheel. Another distant bird of prey glows on the cliff, radiant as the sun sinks; too far away to identify. Perhaps even a Golden Eagle. They're all here.

Tuesday

The pair of mystery hawks has appeared again, this time circling right over the guest house, hanging there almost motionless, staring down at us as though expecting something. I am able to lie on the grass with binoculars and bring them into focus. I can now tell for sure what they are. Bonelli's Eagles. Fabulous birds. Streaked on a white breast, beautifully marked wings and banded tails, with a bolder band at the tip of the tail. They are surprisingly low, apparently comfortable around this place. I text a colleague about my find. "Was sure I'd found a couple of Gos – perfect habitat, lovely day, big oul pale-bellied, streaky hawks hanging on the wind. Turned out to be Bonelli's Eagles. I'm in Spain."

His reply makes me laugh aloud: "I hate it when that happens."

Wednesday

We visit the Cranes at Gallocanta, tens of thousands of them gathered on the snowy plateau, honking, dancing, swirling overhead. It is a glimpse of medieval times back home. King John is known to have hunted Cranes with a Goshawk in Norfolk in the 13th century. Britain is getting its Cranes back too. They have returned to East Anglia under their own steam, and they are being put back in the south-west by conservationists wearing Crane suits to stop the new birds imprinting on people.

MARCH

I've recently become a board director of the journal *British Birds*. We meet at the offices of Natural England in Peterborough, with a grandstand view cathedral-wards. I'm watching pigeons with one eye, thinking I might glimpse a Peregrine. I've been re-reading *The Peregrine*, now complete with Mark Cocker and John Fanshawe's preface and notes on the author's hitherto unpublished diaries.

Baker describes meeting falconers carrying a Goshawk and he later recounts a Peregrine hunting a Sparrowhawk among the trees in a wood. It sounds more like the behaviour of a Gos. I am tempted to think that where he watched, he might have been as likely to find an escaped Goshawk, or a wintering bird or birds arrived from the continent. It reminds me again of Saxon king Byrhtnoth's hawk, released to the wood as a morale-boost for his Saxon army on the eve of the Battle of Maldon, against the invading Danes.

I've discovered someone else bewitched by the image of a Goshawk. Gerald Summers, in *The Lure of the Falcon*, describes discovering, at his Aunt Maud's house, "one picture which may well have had more effect on my mind than I realized. It was a splendid likeness of a goshawk which had just made a pass at and missed a rabbit, whose hindquarters can just be seen disappearing into a burrow. I can see that goshawk to this day, her raised crest and splendid topaz-coloured eyes, her huge yellow foot with its bayonet-sharp talons reaching for the rabbit's scut, and her great rounded wings half-open as she hurled herself against the bank. I am sure that this picture fired me with a love of, and admiration for, all birds of prey which has grown stronger as the years pass by."

The Scottish Parliament has passed legislation that will make it easier to prosecute estates that kill protected birds. 'Vicarious liability' makes landowners responsible for crimes committed by their staff, with properly deterrent penalties now possible. Meanwhile, the first day of spring brings a sobering message from a colleague in Investigations. "You wouldn't believe the photos of Goshawks we've just received. Can't say any more at the moment, but it's gruesome."

APRIL

Warm spring sunshine, and the garden of England comes into its own. Goshawks ought to be up. The first warm days of spring and a stiff southerly breeze has brought the migrant birds flooding back, which is always a joy. No sooner has the blossom appeared on the cherry trees than the Swallows are swooping over the farmyards, Chiffchaffs are calling from the still bare branches of the taller trees, Willow Warblers are trilling from the scrub, and Blackcaps are bubbling somewhere in the undergrowth. The Nightingale is back, unusually early, audible from a great distance in the small hours. A Cuckoo too – another bird that conservationists are very worried about – gave its first call later the same morning. It's travelled all the way from Africa, again. I have made a Barn Owl nest box out of an old oak bedside cabinet, and helped Clive the farmer select a spot for it in his byre. I've started work on a Kestrel box, made from an old commode. It needs a home – preferably a large open building, like a barn, or somewhere reasonably sheltered.

A high-profile case has reached court. A man has been charged with illegally using a cage trap baited with a live pigeon to catch birds of prey in the Peak District. He is alleged to have operated the trap on land owned by the National Trust last spring. The trap was found by my colleagues. It is legal to trap certain crow species under a general licence issued by Natural England. Conditions of setting a trap include visiting it every day and humanely killing any target bird caught, or immediately releasing any protected bird caught.

The accused denies seven charges under the Wildlife and Countryside Act 1981 and the Animal Welfare Act 2010, including using a trap for the purpose of killing or taking wild birds. Four of the charges relate to causing unnecessary suffering and failing to meet the welfare needs of a Pheasant, a Carrion Crow and a Feral Pigeon. The court hears that the RSPB installed surveillance cameras near the trap and captured a man on tape, who they claim is the defendant. Someone wearing a balaclava was also recorded on camera visiting the trap.

A report comes in, from a reservoir, ten miles away from my village. "Goshawk, male east of dam from 1210 to 1221, in the company of both a male Sparrowhawk and Buzzards."

Sunday

A chance to visit bluebell woods on the ridge, to take photographs while the sun shines. The Copper Beech trees are half in leaf, smooth-barked and slender-trunked, spaced widely enough even for Goshawks, with few limbs low down, and their upper limbs reaching for the sky. The bluebells are nearing their peak, engulfing the piles of brash discarded from winter forest operations. Some fallen trunks lie as though in a bath of bluebell stems.

I pass through a tree nursery, a futuristic vision of landscape, Bird Cherry trees in groups of a hundred, neatly spaced, all aglow. There are ornamental cherries too, baby pink, Barbie pink, even. And Yews, and Box bushes, clipped in different shapes. I pick my way through the different departments. Skylarks urge me on unseen, overhead, or some way off. Linnets tinkle. Rooks quibble. I sit to write notes and watch the sky as the day warms steadily, the air hazy, moving, the distance wobbling. Spring is starting to feel heady, insect-laden. Whitethroats announce their arrival, busy, vexatious in the briars. Songflights are attempted, from memory.

A quiet lane. Pied Wagtails sing and dance on the wing overhead, landing on the wires above me, stark two-tone against the blue. I hear an approaching commotion of them, and swallows, and I know that a hawk is nigh. I move ahead quickly to get a clear view past the roadside trees and into view comes the hawk with its irate outriders. A small Sparrowhawk, a hint of the male's rufous chest. A young bird. So often these Sparrowhawks seem to be young birds. They don't live too long, perhaps, in this hard and mishap-prone life. It twitches a wing, like a dog might an ear, irritated. The outriders peel away. It circles overhead, then it's off, towards town. It folds, diagonally stoops shaking off the tale-telling rural birds, the all-seeing Swallows, to attack a garden full of peanut-gorged passerines. It is swallowed by suburbia.

A Kestrel flickers by too, beating steadily. I look up to see another hawk-shaped bird circling. Beyond it another, bigger. Higher up. Not a Buzzard, this time. I know Buzzards instantly now. This one has none of the long, upturned, slightly forward-pointing wings, the fanned tail. It has wings held level. Its tail is closed. It flies in narrow, faster circles, never flapping. It just looks very different to the other birds seen today, in the last few minutes, even. It makes a bee-line south-west, in the direction of another wood. Could it be a female Goshawk? Perhaps having a wing-stretch, a look around, while it's hot, and deemed safe to leave a clutch of eggs?

I follow her as she dwindles to nothing in the haze after about two minutes. Looking back to where she started I notice another hawk. This one is lower, just above the spur of the hill, and the pines that crown it. This bird immediately shows a heavier body, muscle, a protruding head, much heftier than the Sparrowhawk of a few minutes earlier. It circles briskly. I am locked on it, and can hear crows whingeing, just out of vision, not going near. The raptor is framed by their mutterings. It kicks into action; tilts. There are two or three deep beats of its wings, these seeming to bend with the effort, the power as it pushes off, propelling it into a diagonal dive, away from me, towards an unseen target, obscured by trees. It doesn't fold and drop, it doesn't go vertically. It generates its own power, all muscle and sinew, pushing off, not falling or tumbling.

I am as sure as I can be that I've just seen a Goshawk.

I walk quickly in the direction it dived. It takes me a few minutes to get there. Jackdaws and Rooks are hawking for insects over the common, in a feeding frenzy. I think these might have been the intended target. Absorbed, preoccupied, flapping, twisting, swooping. It reminds me of the scene in Butch Cassidy, with the Hole in the Wall Gang scrabbling to grab the cash falling from the sky after they've blown up the payroll train, greed over-riding any instinct for self-preservation. 'Think you use enough dynamite there, Butch?' drawls Sundance, smirking, before he notices the posse on their tail.

This place isn't far from what I've come to know as Gos wood ... very close, in fact. That's the wood where I found the Buzzard feather. It has no real name, on the map. So, assuming I am right about these birds, which is an assumption, what might a pair of Goshawks have been doing here? Might a female come off the nest for exercise, a look round, a bond with her mate, give him a little encouragement?

I check with Mick for answers. "Unlikely to get female up in the air once she is on the nest so this might be a bird put off, or a late breeder or non-breeder,' he says. But there is some scope for optimism: 'Hot day though, so maybe incubation gets uncomfortable and soaring enables cooling off."

"Males hunting hard at present so seeing the odd one. Most birds here now approaching the point of laying. Soaring raptors here all day yesterday but no Gos up in the air. Checking some of the black holes in the Gos population with little success. Plenty of crow traps operating now so need to get around a wide area."

Thursday

Goshawks have reached Brussels. A first pair has raised five young in a tiny spruce with an 8-centimetre diameter trunk on the edge of the city, and have been hunting racing pigeons from a great height. I correspond with Rob Bijlsma about it. "Here in the Netherlands, Goshawks are still hesitant whether or not entering the city proper," he says, "although you might argue that the entire country is a city, with some parks. In that case, all our Goshawks are urbanites."

I also tip-off friend and colleague Ariel Brunner who works for BirdLife in Brussels. He tells me he's recently seen a Goshawk chasing Woodpigeons in a Belgian wood. He also mentions the one he saw back home in Italy. It plucked a Garganey off the water at Trieste on the Adriatic, amid great mobbing by gulls. According to Ariel the Gos has now reclaimed the forest north of Milan, and on the coast of Tuscany.

Friday

Easter weekend. The sky is cloudless. I venture to the garden to sit in a deck-chair with the biography of T. H. White. It's the height of the afternoon and most things are feeling a bit wilted. It's hot up there. I've got ponds receding around me, tree roots running out of credit, water butts dwindling. And it's only April. We haven't seen a shower all month. Then the Blue Tit gives the alarm, from a sentry post behind me high in the tangle of holly, elder, juniper, Honeysuckle and beech.

'*Cheeee cheee chee-chee-chee-chee-chee.*'

Hawk?

I look up. Nothing. Seconds pass. Still nothing. Then something. High above me and the tell-tale Blue Tit, dark against the blue, is a falling shape. Not a bird shape. A kind of heart shape. Going across the sky, gently – not steeply – downward. What on earth is that, I'm thinking. It's disguising all its salient points: head withdrawn, wings completely tucked. Its not in a hurry. It's coasting. Joy riding. It's biggish too.

Why did the Blue Tit yell hawk?

Now I'm thinking Goshawk. Really thinking Goshawk. I lose it behind the house roof. I'm on my feet and running, dropping T. H. White at the back door and scrambling towards the front door. I get on the bird again as soon as I reach the garden path. Now it's circling, but heading away, not very high. It is more business-like than a Buzzard, its turning circles are smaller, its demeanour less sluggish. Buzzard is a bird of leisure. You've only got the Buzzard's word for it that it ever catches anything with a pulse. This bird means business, a little bit of urgency about it. *Oomph.*

I watch it till it dwindles. It goes left of Muffin Wood, flapping once or twice, wings wobbling, to generate a bit of extra pace, maybe. An unseen Buzzard cries a few times, sounding a little feeble. I take it to be the male on sentry duty at Muffin or even neighbouring Esme Wood, tipping off his mate on the nest. But he doesn't come up, to make proper sense for me of this interloper, give me something to scale and judge it against. It leaves me guessing. I hobble back across the road and into the house to grab binoculars, and come back to scan. I quickly pick up a Buzzard over Muffin Wood – an obvious Buzzard – but not, I am certain, the bird I was watching before. This is the female off the nest in Muffin, I think. Reasserting her territory.

The Blue Tit said it was a hawk, didn't it? The signature tune, the entrance fanfare, for the hunting hawk. Blue Tits don't give off for any old displaying Buzzard, do they?

I grab a notebook, close some doors, head out again, to sit on the edge of the field opposite and scan over towards Esme. A Buzzard is up there too now. Perhaps the same female.

I walk the field margin on tyre tracks. The desert-hot sun beats down, merciless. I pass through wafts of hawthorn blossom, drowsy and seductive. Birds are quiet in the heat. I duck under branches of Field Maple and into the interior of the oasis, a clump of trees and scrub around the pond that is the last remnant of the former Great Marsh. The parched ground is dappled, sprinkled with blossom petals, tinted blue with bugle, the odd pigeon feather. Nothing suspicious. No Moorhen croaks. A Robin lilts dreamily. A distant lark song filters through, delivered as though from memory, against the haze.

I wonder about the merits of my description of the hawk I've just seen. This business of getting it right, or not. Like there is a code, one I don't yet fully know. Perhaps it's just a test of being able to describe how it felt, how you felt. A test of being bothered to try. To show that you want it enough. To earn it.

My village is midway between where I thought I saw the Gos last week to the south, and the gravel pits where others have reported one in recent times.

I sit down at the junction of two ditches, three fields, where the hedge peters out among sweet-smelling grasses – more remnants of the marsh, the drains that bled it. Muffin Wood shimmers in the heat, two fields away, with half-dressed oaks. I scan for Buzzard sentries, or who knows what might be perched on those bare antennae branches. Nothing. Not even a lark now, over the hundreds of acres around me: green wheat, oilseed rape, crops already thick as doormats, uniform as Astroturf.

At my approach, a Buzzard cries from the depths of Esme. Two flap out, together, and away, back in the direction of the house, low and hard, then up and drifting, calling. Complaining about me, maybe. I enter the wood, wincing as I negotiate a dry ditch, torn calf muscle painful on these slopes, these tricky jump-steps. I can't disguise my progress. It would be less noisy to paddle in dry cornflakes. A Blue Tit does its whistle-blowing again – for me? Or for a departing Sparrowhawk, unseen by me?

I don't stay long in the wood. I am making too much noise. I don't find anything. I never expected to. I sit on the edge of it, on the iron fence, looking out across the parkland, dotted with Rooks forking the turf for invertebrates, broods raucous in the tree-tops beyond. Rich pickings for Goshawk, a pick 'n' mix store, a smash-and-grab, ram-raidable kind of a place. A Mistle Thrush rattles, Jackdaws creak, high in the chestnuts. A woman passes behind me, walking her dog on a long lead. The dog sees me, but I don't think she does.

On the street again, walking home, a neighbour calls across to me. "Have you been in the estate?"

"Yes," I call back. I walk across.

"Do you know what the hawks are that circle over the wood?" she asks me. It takes me a while to get my answer in, as she goes into a long spiel of further detail.

"Buzzards or Sparrowhawks," I interject, spotting a gap. I don't bring up the subject of Goshawks. Might be complicating things too much. But who knows what she has really seen.

I've checked with local birders, for any possible corroboration of my 'Gos'. "Would that correspond with any other local sightings?"

"Not aware of any. Please feel free to write a description of yours," says Richard Bashford. I think better of it.

"Afraid not, Conor, but at this time of year, I wouldn't really expect any," says Darren Oakley-Martin. "I'll let you know if I hear of anything on the grapevine ..." he says, adding: "I'm thinking you probably have more recent experience of Gos than anyone?"

"You flatter me," I tell him. "A Blue Tit told me it was a hawk ..."

"They are rarely wrong!" he replies. "Unlike Swallows, which will alarm call at a passing cloud ..."

I run it past Mick. "Would Gos fold and dive slowly like this, over open country? It looked more like a display dive than an attack – not that fast or steep. I know Buzzards dive as part of display."

"Yes, definitely," he replies. "In open country Goshawks stoop like Peregrines. I have seen them strike Feral Pigeon and grouse using this method. The grouse was hit but not killed, so was chased on the ground. Blue Tit alarm is a great give-away. Hope you see more of this bird."

Rob Bijlsma gives his view. "Goshawks often hunt while stooping in a Peregrine's fashion. How else to outpace racing pigeons! But remember, even Sparrowhawks may do so. The enclosed paper gives an account of what the brute is capable of."

No bird book I've yet come across makes any reference to this, and no descriptions of Goshawk ever suggest you might see this hunting behaviour.

* * *

It causes me to reflect how, two years ago, one May evening, I saw what I took to be a Peregrine, near dusk, cruising high, then stooping at and entering a very small copse, by the airfield. There was a tremendous commotion of panicking Pheasants, and while I waited for some minutes to see any outcome, it all went quiet, and no birds emerged. It is only now that I wonder if it was a male Goshawk. Would a Peregrine go for game within trees? It seems unlikely. And certainly not to provision a nest at least five miles away, light fading.

I can at least find a Goshawk on prime-time TV, as the BBC show greatly decelerated footage of Ellie the Goshawk to illustrate how these big birds can squeeze through narrow apertures, even at high speed. Presenter Mike Dilger makes the mistake of waving his hand in front of Ellie. The bird is clearly unfulfilled by the lure they've used to get her to perform. In an instant she leaps from the ground and nearly removes his finger.

Monday

Nigel, an ex-pat, has written to me from Berlin. 'I have read your super article, when checking for information on Goshawks in the UK. Could not really identify your park in Berlin ... but many locals and NABU (RSPB partner in Germany) are pretty secretive about nest locations, even though the locals either know their birds or maintain a typical Berlin nonchalance.

"Anyway, we live in south Berlin away from the big city action. Before our flat is a small natural lake which is a small oasis for wildlife, and from our balcony we have spotted: foxes, martens, squirrels, and of course loads of birds. But about 10 minutes by foot is a small park constructed of rubble from the war. It is reasonably high for flat Berlin and on top is an observatory. Here the Goshawks have a massive nest in a spruce. I'm no bird expert and had never seen a Goshawk before coming to Berlin. I used to spot this bird whilst jogging, unconcerned with humans. One local tells me that the female has nested for eight years and last year had four chicks.

"The nest is supposedly watched, but what is typical 'Berliner attitude' is that life goes on around the birds, with joggers, dog walkers, kids maybe watching or ignoring them. In March I watched them pair on a branch about 20 feet away. A wonderful sight just to see the difference in size ... and all in the middle of a big city. Birdwatching for lazy guys! Some old man watched a bit with me, but walked on without a word.

"Other people are fascinated and dog walkers usually ask what the bird is ... it is usually the female sat around, imposing and majestic, and screeching like some crazy Herring Gull. Last month I heard her from about half mile away.

"One of the 'experts' says that they hunt in the cemetery opposite and she is partial to squirrels. I have found mostly bird remains ... could only identify Magpie. Yesterday one was near our wooded lake but was swooped on by two Hooded Crows quicker than strike jets.

"The female killed a Mallard just below our balcony early one morning. Neighbours were woken up by the noise and, finding 'the body', blamed the Fox. Sure enough, one hour later she sat in a high tree to view her kill. We were watching with coffee on the balcony! The bird just bided her time and then swooped and picked up the duck in one grab, leaving only orange legs on the path. She sat on the limb with it for about ten minutes and then flew off between the trees, dodging agitated crows. The Hooded Crows mob her like crazy.

"Someone reckons the female ate her mate during the hard winter just gone. He also said the bird has visited his apartment block and consumed a pigeon on the balcony."

***

Mel Worman, another ex-pat who lived in Berlin for a time, has also been in touch. "Like you, I'd like an answer to the lack of Goshawks here, and other wildlife in our cities,' she writes. 'And like you, I think Berlin inspired me to try and find out more. But unlike you, I'm not sure how yet! Thanks again for reminding me of a wonderful bird, and a marvellous city."

***

There's a pressure that comes with the last day of the Easter bank holiday weekend being beautiful and sunny, in knowing how to make the best possible use of it. I struggle with the conflicting demands of needing to sit still and write, do odd jobs, get out into the blue yonder, and do things like eat.

I get the local map out to look at woodlands again. I've added my own Goshawk sightings (possible and probable) as coloured stickers, to go with those of other people, further to the north. It occurs to me that there may be as much chance of any prospecting Goshawk erring in favour of a small isolated spinney somewhere, one that the public can't ordinarily visit, as one of the larger, but still barely adequate, more obvious places. The path of least resistance to the wandering young Goshawk looking to settle. I check to see if there are any tracks to the copses on the ridge to the north.

I roll out on the bike in the middle of the afternoon, the sun high. The paddock is overgrown with Cow Parsley, the odd thorn coming through. The Grasshopper Warbler is reeling again, on, off, on on on on, off. It flits into the hedge, down into the long grass, up on to the barbed wire. Through the binoculars I can see it turning its head, beak slightly agape, making the sound, although this is faint against the sound of the breeze, the chirrup of sparrows around it. It shows a yellowish tinge in the strong sunshine. A little triumph, and a little piece of the former Great Marsh, back amongst us.

Watching a pair of Buzzards I can't helping thinking that one of them appears to be almost 'enjoying' the attentions of eight Rooks, as it leads them ever more heavenward in their futile fury. Could there be an advantage for the Buzzards in the Rooks being side-tracked in this way, away from nest-guarding duty? Might make it easier for partner Buzzard to do a quick raid ...

I wonder about the differing behaviour of corvids towards Goshawks. I've heard of Goshawks capturing crows that have been pestering them. The mob has dwindled to two by the time I lose them against the bright sky.

Corvids have the reputation of intelligence, but so often their mobbing behaviour looks like a waste of energy – a 'locked-on' reaction rather than a sensible one. This Buzzard even performs a triple salchow at one point, spinning sideways three times at one point, which leads me to conclude it is showboating, or taking the mickey.

Emerging from this small wood the view makes me breathe deep – like seeing it for the first time. Extensive to the south and west, the scrub is growing well on Hungry Hill, hawthorns knee-high. Linnets singing and twanging and wheezing, breeze warm off the slope. Jays squall in the wood over to the right. Magpies work the field margins like bargain hunters. It doesn't feel like there are Goshawks around here. I push through the hawthorn hedge where there is hint of a gap. The Buzzard takes off again from the south edge of the wood, and crows go ballistic as it passes them. The view to the east is good too. I can see where I cycled yesterday, over on the other ridge, forming part of the rim of the great bowl that has the airfield as its centrepiece. Even the gigantic crop fields look OK in this context, and the landscape to the west, with our house, our road, in its midst, suddenly looks like 90 per cent woodland, from this lower-than-you-think angle.

If there are Goshawks around, this is the place to sit, with a powerful scope to scan above the woods from miles around for displaying birds in the early spring. The Whitethroat buzzes me again from the hawthorn hedge. I do a 'pissht' sound and it erupts from the hawthorn's crown into a spectacular song flight, like a little Orca showing off in a sea life centre. It drops again, singing all the while as it disappears into another part of the hedge. I can just make it out as it hides among the florets. To cap it all, one of the five Linnets I can see swooping around lands on an exposed perch in the Elm, and sings sweetly, rosy breast aflame in the sun.

There is something especially meditative about warm sunshine and a steady breeze, and this time of year. Something carefree about days like these. I could sit on Hungry Hill for hours. It reminds me of the view from Witches' Craig near Stirling, not because it looks the same, which of course it doesn't, but because it evokes similar feelings. Amazing how something that from down there barely looks like a hill at all can give such an aerial perspective on the world below, the world I move in. It's a hawk's view, of a place worth living in, nesting in, if a quiet corner can just be found. I am happy that I am looking at a Goshawk's world now. I just can't be sure they've quite got here yet, to breed. But I hope I'm wrong. They could be down there, or over there, somewhere.

Wednesday

It's 25 years since the spring morning when Europe awoke to news of the Chernobyl nuclear accident in Ukraine. I have dug out my old diary from the time; a postcard from another era. I was a student then, at the end of my second year. The continuing struggle in Japan to cope with the aftermath of the Fukushima nuclear plant's failure to withstand an earthquake and tsunami adds further piquancy to the retrospectives. I now have the chance to see the Chernobyl exclusion zone for myself.

Chernobyl is close to Ukraine's border with Belarus, which means white Russia. Belarus is the westernmost of the former Soviet states. Its capital Minsk, with its turbulent, tragic history, lies on the road from Berlin to Moscow. The Nazis flattened it on their march to and their flight back from the Russian capital in World War II. It has been a pressure point for fierce and bloody conflict throughout history.

Today, in peacetime, Belarus is of global importance for its vast, carbon-storing forests and wetlands, and some rare species that still find refuge here. The RSPB has been helping to set up a non-governmental nature conservation organisation here, a pioneering move in a country in which not much is non-governmental. Like anywhere else, non-state bodies can do things that governments cannot do, as well as will not do, or do well, and that includes attracting funding from outside the country.

Meetings over, there is a chance to travel south for several hours on a very long, straight road through the wide open flatlands and forest blocks to the south of the country. I am part of a group being shown how the maintenance of vast peatlands with adequate water levels benefits biodiversity, carbon storage and prevents fires. Besides its other harmful impacts, fire can release radioactive contamination.

I've joined a helicopter flight with my hosts over the special reserve established after the Chernobyl disaster. Much of the fall-out drifted north into Belarus, and beyond. It has been a tragic story, but the reserve gives a fascinating glimpse of how nature can regenerate. The experience is dream-like, cocooned in the bubble of the chopper, locked in by ear protectors against its whirring blades, face pressed downwards as it tilts. Below us is as close to true wilderness as can be found in central Europe. Around the abandoned farms and villages forests thrive, and the sprawling, sparkling river spreads itself out in ribbons and floodplains. There are wide reedbeds with browsing Elk sploshing noiseless in water up to their waists. Herds of deer scatter, Wild Boar forage, eagles glide over meadows. There are Goshawks down there too, though of course we don't see them. It is a privileged insight into nature's powers of recovery, albeit in the context of a human tragedy. On the southern horizon the shadow of the now derelict nuclear plant is just visible beyond the border, which we cannot cross.

I also have a rare chance to see for myself the last and best substantial fragment of the ancient forest landscape that once covered much of the European plain. Archaeology experts like Derek Yalden and Umberto Albarella have used its example to estimate what Britain might have looked like once upon a time – 7,000 years ago, for example – and what our wildlife might have been like.

I arrive late in the day at the forest, the Bialowiecza National Park announcing itself with two huge, rearing white stag sculptures, on opposite sides of the wide road, in front of the wall of pines. Viktar, host, guide and driver, takes our party to our accommodation deep in the forest. Like much of Belarussian architecture this is a modern building, a brightly coloured hotel complex with a new, synthetic, unused feel, contrasting with a backdrop of dense, towering trees. There are very few people around today.

The air is cool in the forest, with a fresh hint of damp, of snow, even. It feels more like mid-March than late April. The hotel doorway is unusually tall, elegant, like in a dream. It opens onto a long corridor, and a red carpet that leads 50 metres to a wide desk at the far end, behind which sits a female receptionist, straight-backed, facing this way, as though in expectation of our – and only our – arrival. It adds to the surreal, dream-like feel of the place, the peculiar disconnection between the interior and the wild forest just outside these walls.

I waste little time in getting back outside to explore in what remains of the still evening, the filtered light. I take a wide track, which has been comprehensively torn up by Wild Boar tusks. It is as though a gang has gone on the rampage with hoes and picks. A Tree Pipit trills vigorously from a pine on the edge of a clearing. Warblers call and a Song Thrush pipes; Cuckoo and Hoopoe calls carry across the landscape. I am on the edge of a grassy meadow and wetland, with Roe Deer browsing quietly at the far end. I realise a large raptor is drifting slowly across the sky, low over the trees. Perhaps a Spotted Eagle.

Viktar employs a researcher who finds and monitors Spotted Eagle nests in Belarussian forests. As it happens he finds and records nests of Goshawks at the same time, as well as nests of Great Grey Owl and Ural Owl, and the occasional forest-dwelling Eagle Owl. The latter often nest on the raised root-ball systems of Alders, in the swampy parts of the forest. There are several of the relatively tiny owl species here too, like the Pygmy Owl.

Whatever I saw this evening it was too languid for Goshawk. Even here, where people are blasé about them, I know my chances of finding the Goshawk are limited, on a brief visit. There is of course much else to see, to be alert for.

Progress becomes difficult in the thick under-storey. In the dwindling light I opt in the end to cheat a little, and visit the enclosures where Bison are kept as part of a captive breeding programme. Bison were reintroduced to the forest after being hunted to extinction by the occupying German army in the Great War. It is said that when the Polish army re-took the forest in 1919, after three years of exploitation and railway construction by the Germans, to help extract timber, the last Bison had been killed just a month earlier. By 1923 Bison survived only in zoos in other parts of the world. These were used to replenish the population. Today there are about 300 living wild in the Belarussian share of the forest. The Polish population is larger, but the two are separated by the border fences that splice the forest, with a strip of no man's land, including a wide belt of trees, between the two countries.

Based on present-day numbers of the Goshawk here in this forest, Yalden and Albarella have estimated that there were about 14,000 pairs of Goshawks in that Mesolithic forested landscape of primitive Britain. The Goshawk was, if not the alpha predator in that food chain, or ecosystem, then certainly one of them. It is better suited than most species to a landscape mosaic of mature forest, with open areas of fen and moor, and where these trees bordered open ground.

Contrary to popular myth, the landscape before we turned up wasn't all forest. About 60 per cent was tree-covered, the experts reckon, based on pollen analysis. Allowing that some trees leave more pollen traces than others, and than non-trees, and given that other factors have always been at work, even before humans, to prevent trees taking over entirely (fire, wind, browsing animals, disease, changing water levels, freezing weather), the other 40 per cent was left open in some way. The highest ground was probably a bit like the scraps of intact moorland we have today, the wet areas were fens, and there would have been regular gaps between the trees.

At first light we walk from the hotel through a deserted car park to Viktar's truck. The air is ringing with familiar birdsongs mixed with some that are less familiar: Serins, Chaffinches (by far the most numerous species here – an extraordinary 150 pairs per square kilometre), Chiffchaffs, Hoopoes, Great Tits, Blackbirds, Robins, Greenfinches.

Thursday

Early in the morning we drive to the border-crossing checkpoint. In my mind's eye I've pictured an open place, a wide space, cold and exposed. Instead, the gate in the fence is ringed with more gigantic forest trees, close-packed, like a little clearing, an open-air theatre. We park the car and wait. A soldier appears on the road from the direction we have come, hurrying without running. A young man, in full uniform, he catches us up and talks to Viktar. He unlocks the gate. We walk across the short strip of raked sand with our baggage. He asks us to walk in file, to minimise disruption to the sand, which he'll have to rake again after us. Apparently this isn't as simple as it looks, nor the task as welcome as you might think for someone who may see just a few border crossings a week in the shoulder months of the season. Words are cunningly inscribed in the sand, to make it difficult for anyone sneaking over to rake the sand behind them as they go, to cover their tracks, goes the thinking.

The forest between the fences, beyond the strip of sand, must be some of the best preserved, most undisturbed of all, a strip that follows the length of the border.

We arrive at the office of the border guards as the gates are locked behind us. The guards take our passports and our immigration forms, and disappear. I take the opportunity to watch the sky, beyond and between the conifers looming around the edge of this peculiar scene, of modern brick buildings, barriers, signs and car parking areas, all empty.

A raptor appears from the direction we have come, the east, rising vertically into the blue, from behind the trees, peaks, and descends again. I point and blurt out 'look!' conscious that decorum has to be observed here. But the bird has gone.

It reappears, on one final peak of the rollercoaster ride it is describing across the sky. Goshawk? Viktar nods. "Yes I think so ... Goshawk!" he laughs. He may be telling me what he thinks I want to hear. I'm not sure. Could have been another Buzzard – the fool's Goshawk. Crossing the border in style, unconstrained.

The show isn't over. I notice another large bird in the sky – higher up. Much higher. Circling. I get Viktar on to it. He focuses his binoculars, as I do. It's a White Stork. But then we both notice the hawk shape even higher than the Stork. A much smaller bird, of course, but not that much smaller. A Sparrowhawk or Goshawk, for sure. They move across the sky, over the border, and when the formalities are complete, we are able to follow them.

The lack of physical barriers in the landscape here is striking, both in this country and later in Poland, Very few fences and walls, and no hedgerows in the farmed countryside. You are free to walk where you like, in fields and forests. No one will challenge you, if you are doing no damage. And much of the land is privately owned in Poland. Communism didn't dispossess private owners here, the way it did elsewhere. You can walk in crop fields here as long as you aren't damaging them. No one is waiting to harass you. Almost any woods are free to explore. People here harvest mushrooms and forest fruits at will. Some locals sell what they find, from little stalls by the roadsides in autumn.

But at that moment, on that border crossing, the Goshawk for those few seconds stood for all birds, all freedoms of movement, and any of both of these things that we may sometimes take for granted.

MAY

April has been the driest since T. H. White lived in his keeper's cottage, in the late 1930s. It hasn't rained in the village all month, and hardly in March either. I notice frazzled tips on the vine over the cabin in the back garden. Even the long-established plants no longer have their toes in the water table. The farmer has no vegetables to sell at his gate.

I feel a little disoriented – out of step with the spring, having been following it so closely until the time away. Slightly lost track of it, lost the rhythm. I sit under the cherry tree and listen to it sway, brushing away its debris forming a layer on the canvas of the hammock. The blossoms are now past, prematurely in many cases. Jaded and dried by lack of moisture, shed early as economies. The spring is turning to summer, walls of leaf green and shady places, contrasting with the lights and shadows of a fortnight ago, and those of eastern Europe, where spring is several weeks behind our own.

Rob Hume, expert eyes and ears in the New Forest, has sent me news. "Good hour this morning in the Forest. Honey Buzzard gently chased immature Goshawk; Goshawk later chased Honey Buzzard, equally innocuously. Honey Buzzard displayed once then flew away. Goshawk was chased by a Hobby, while a Peregrine flew by underneath!

"This Goshawk four times soared up very high, tilted up its tail and pulled in its wings, like a giant Meadow Pipit, tipped over head-down, closed its wings, stuck out a leg like a clenched fist, and simply plunged headlong out of sight into the wood at fantastic speed. Each time it reappeared within two to three minutes."

***

Sunday

There aren't many Sunday mornings in May, in a lifetime. I can't bear to miss them. I get up soon after dawn. There are puddles on the flat roof. The rain gods have answered our prayers, though not yet with the prolonged downpour that's really needed. I look at the map again, and resolve to cycle where the Goshawk went, check out the small spinneys. A long shot, but maybe it could be in there.

Two weeks have passed since the possible Goshawk scorched across the sky over the garden, the house, that afternoon. I have checked its route on the map, to see where it might have been headed as it dissolved into distance after I watched it circle.

The House Martins are back at last, in modest numbers. And the hawthorns in the front garden hedge are filling out. More rain has been promised. I link the water butts to the pond to top it up. The Swallows tell me to look up, and sure enough a Hobby is cruising overhead, never too far behind them.

JULY

I've been making preparations for another trip north, a bit later than I would have liked, to catch the tailend of the nesting season. Mick has provided an update: "Gos are suffering a bit with bad weather, so brood sizes are down and the latest attempts have mostly failed. Nevertheless, should be some young still on site. The earliest fledged birds will be dispersing now."

Tuesday

On the road north I stop off in Perth, for a long awaited pilgrimage to its elegant museum, with its Roman pillars, to find the last Goshawk pair that survived in Britain before their death spelled extinction for the species here, in 1883. I make my way through to the natural history collection, past the spectacular array of oil paintings, vases and armour. The nature exhibits are set up as a tour through geological and evolutionary history, through to the most familiar of modern-day Highland wildlife.

I take my time, enjoying the artefacts, knowing there will be Goshawks, perhaps the most celebrated pair of all, somewhere towards the final chapter of this story. Everything is here – all the usual suspects – Golden Eagle, Wildcat, Pine Marten, Capercaillie ... The Osprey pair has been reconstructed along with eggs and chicks at the nest, with a note card explaining where the specimens came from. I say everything is here, and it's all beautifully done. But I'm running out of cases to inspect, and I've looked hard at the forest one, and I can see no Goshawks. Perhaps they have a special plinth of their own, elsewhere.

Much as I have wanted to find at least these ones on my own, I realise I'll have to solicit help. I go to the main desk and explain my mission. The staff are friendly and helpful. It turns out the fabled (at least in my mind) Goshawks are not on display any more, although they are still held in the archive. I would normally be able to meet them, but the man who curates the natural history artefacts is on leave.

I can see the ironic side of failing even to find these stuffed and mounted birds, although this is tested when I return to the car to be greeted by a parking fine. I hold to the hope that a pleading letter to the council explaining my delayed and fruitless search will be treated sympathetically. This hope will later prove forlorn.

**

Three hours later I reach Mick's in the north-east. We are visiting a bit later in the season than we might, and young hawks are less attached to their natal site by the day, but, true to his promise, Mick 'whistles up' some fledgling Goshawks for me. His calls are answered, from deep within the forest. He has a key for the tree trunk barrier that excludes all but the most energetic walker from this vast state forest. It's quite feasible that these young hawks have never before encountered a human being. It's evening. We walk in that direction, over mossy boulders by the bed of a stream, past scattered prey remains.

There is movement in the canopy, and an adult Gos arrives on the scene, and I don't know if it is in response to Mick's whistle, but it is carrying prey. Seeing us, it veers away at right angles, and off again, through the trunks and branches, vanishing within seconds. As it turned, it may have brushed a twig, encumbered, and a feather is falling. It settles gently on a tree stump, as though specially presented that way. I go to look, expecting it to be from the bundle in the Gos's clutches. But in fact it's a Gos breast feather, pale and lightly barred.

Wednesday

We visit three more sites, with youngsters dispersed but still calling, nests successful. I even find one or two of them. I ask Mick how often, in the four decades he's been monitoring the slow but steady progress of Goshawks in re-colonising this region, he's seen the birds hunting. The astonishing answer is "just twice". One of those occasions was a strike at Jackdaws in his back garden (the crows scattered, the hawk appeared and disappeared behind them, and he later found a scattering of black feathers). The other was a bird 'perch hunting' down the gulley near his house. He didn't see an attack, just the hawk flying, perching, looking, flying again.

The last nest site we visit is now a private woodland, and the birds have chosen a low tree right beside an off-road vehicle rallying track. If this is evidence of a bold bird, this is backed up when the young male escorts us away from the premises; calling and flying around above us for some distance as we beat a dignified retreat.

Thursday

On the road south once more I pause again in Perthshire, for a visit to Birnam, where that last pair of Goshawks had attempted to breed before being shot. Birnam is best known as a location in Macbeth. Looking again at 'the Scots Play', as my actor friends insist on calling it, I note that Shakespeare via Macduff talks about a "hell kite" – which I am sure is a reference to the Goshawk. There would of course be huge consolation in finding a pair here alive, today. So I have good reason to curse the weather – which is closing in rapidly – and the lack of time to search.

I have another homage to pay, at a nearby estate called Murthly, birthplace and family home of Winnifred, who became the Duchess of Portland. She was the first President of the Society for the Protection of Birds (later Royal), a post she held from 1889 until her death in 1954. She gained renown not only because of her compassion for animals, but also her work for the mining communities of Nottinghamshire. It's curiously difficult to find out much more about her.

I park up and cycle a long avenue to the gates of the estate. Murthly Castle is a beautiful setting, in the wooded valley; peaceful, with no cars allowed. It has turned into a typically murky, heavy Scottish summer evening, atmospheric, with moisture hanging over the wide Tay, the summits of the wooded hills obscured. The castle is tucked away behind some massive trees, a romantic, turreted stone building. I explore the chapel, and find a cluster of Stinkhorn fungi glistening beneath nearby trees. The air is thick with their pungent scent.

Fresh from my latest training with Mick I issue my best Goshawk whistle, which has the effect of bringing in a Sparrowhawk for a brief and possibly dismissive look at me, and then gets the Buzzards going. A hungry-sounding young bird and some upset parents give me 'pelters' from the top of a Secoia. I find Pheasants in a pen, fed by several barrel 'hoppers'. It isn't hard to imagine a child here developing a great love of nature. I wonder what Winnifred knew of Goshawks, or if she heard the news of their final extinction. Had she done I'm sure it would have been with deep regret.

Surely there must be Goshawks here or near here nowadays, but the latest bird report for the county suggests not. The forested hills and valleys of the shire would be challenging to survey.

The rest of the journey back to Mum's in the Borders is done in Scotch mist and spray. I stop to examine a pulped raptor on the road near Hawick. I suspect it's a juvenile Buzzard, but have the wing drying off for further inspection in daylight tomorrow.

A quick note to Mick. "Really great day in the forest – so thanks again. Learned loads and hope to make good use of it. By the way, we were blessed with the weather – the Forth Valley this evening was dreich!"

"Enjoyed your visit, good discussion on issues," he replies. "Visit again sometime."

AUGUST

Fifty years ago, in the autumn of 1963, T. H. White embarked on a lecture tour in America. He was by this stage a highly successful author, and in huge demand. He had made it big when his now classic historical fantasy *The Sword in the Stone* was chosen as Book of the Month in the US. With this tale now adapted for the stage, he spoke about his life and work to huge audiences on an exhausting itinerary. On the return journey from the US his ship visited ports in the Mediterranean, including Piraeus, Athens. He had been very keen to explore the city. 'It will give me a chance to photograph some Hadrian buildings,' he had written in a recent letter, the Roman Emperor Hadrian being a particular enthusiasm.

When the *SS Exeter* docked there on 17 January 1964 White was found slumped on the floor of his cabin, dead. The verdict: acute coronary heart failure. He was just 57. With no close family, he was laid to rest in a quiet corner of an Athenian cemetery. Despite the enduring popularity and success of his books, some since made into Disney classics, White has been in many respects forgotten.

As fate would have it, I have found myself in Athens, for meetings with our BirdLife representatives here, and the Leventis Foundation. Greece is close to the south-eastern limit of the Goshawk's European range. They aren't often recorded near Athens, although birds do drift around the Mediterranean islands. There's a distinct subspecies – called *arigonnii* – of Goshawks on Corsica (France) and Sardinia (Italy), smaller, darker birds, adapted to life in the dense forests of the steep mountainsides: a bit like an avian resistance movement, near impossible to find there. Believe me, I have looked in both places. Foundation chairman Tasso Leventis very kindly sent me a sequence of remarkable photographs he took in Cyprus of a Goshawk nest. He was able to set up a hide on a steep hillside, allowing a view at eye-level across to the nest in a pine.

With business concluded, I have a chance to explore this great city, now fallen on troubled economic times. I pass the Akropolis and the fleamarket, but my main interest is in exploring what is called the First Cemetery. Somewhere in its midst lies the unscheduled resident, Tim White.

I venture in, beginning in the bottom corner, thinking to be systematic. I move through glare and shade, under crackling pines and buzzing cypresses, among a shantytown of marble edifices, disconcertingly life-like busts, portrait photos in frames, lit candles, stray oranges. The place pulses with life: warblers scold, doves croon, cicadas scratch on papery bark. It is a village of mausoleums, monuments to love and memory, piled in and around wild nature. Some of these edifices have steps, cellars, doorways, windows. A marble teacher sits with knitted hands behind a marble desk. There are presidents here too. I walk the wide pathways between, there is order here, a grid-iron pattern within the forest.

I begin to feel like Tuco searching the graveyard in *The Good, the Bad and the Ugly*, looking for Stanton's grave. Crickets rasp, Goldfinches tinkle, Flycatchers tick, Swallows thread their way between the trunks and slabs. Magpies and Hoopoes merge black and white with the dappled whole, playing hide and seek. Somewhere out there is an Englishman, largely forgotten. He wrote a book called *England Have My Bones*. This was his breakthrough as an author, gave him the confidence and wherewithal to pack in private school and go feral with his Goshawk. It is not lost on me that England doesn't have his bones, seemed not to want them. Yet Athens, here, of all places, does. Nowhere could feel more different to the graveyard of an English country church, to that dim, sleety Sunday afternoon in November, when I looked for his cottage.

What am I actually doing here? Perhaps the grafter in me wants to sweat for it, to have to work. The logical me wants to find it by a process of elimination, systematically. And the me that wants to be lucky, that wants to trust fate to see me right, the unerring hand of infinite wisdom to guide and smile upon me, to do something a little spooky, make me feel wonder, and blessedness, wants to chance upon it. And realising just how vast this place is, I begin to meander, to intuit, seeking shade between pools of hot light, prettier paths, following the Hoopoes, that they might lead me there, give me some sign to read too much into, give me a better story. I remember Mick talking about the theory of 'optimal foraging', coined by someone called Levy. It seems random but it's as good as any. It's what other animals do. It works for them. I sit on a slab and look again at the biography. There's a note that says Tim was buried in view of Hadrian's Arch. Perhaps I should be heading uphill. I even entertain the thought that his grave isn't here anymore; has somehow been lost.

After a couple of hours of such midday Mediterranean sun molten musings, and in the kind of reverie reserved for northerners in such conditions, it has long since become clear to me that for all the 'it could be you' wishfulness of the project, I won't find Tim this way, even though I'm sure his modest grave could stand out well among these shrines. I would have as much chance of finding a Goshawk here. I will need help.

Clearly, I need to do more research. I find an Internet cafe. Online, I turn up an article in an Irish newspaper. 'Lonely Man White' it says. It has a better photo of the grave, showing a bit more of the cemetery around it, including what looks like a small brick outbuilding behind, with shuttered windows. That gives me a search focus.

Sunday

Early morning, armed with this new knowledge I try again, and fail again. I only give it an hour this time. I have all but given up by now, grown accustomed to this sense of not finding. A woman passes, crow-black hair matching her dress. Two older ladies follow. She has some English. I show her the book, the grave. The older ladies usher us all into some shade. Although it is early, already the air is heating up.

"Who is he for you?" asks the lady in black, looking up at me from the book cover.

"He is a writer," I reply.

Her face now illuminates, relieved. Someone at last looks happy, in this city of so many anxious faces.

"What is the name?"

"T. H. White."

More discussion follows, now there is nodding, and smiles, and discussion in Greek.

"Go with him," she instructs me, nodding and waving towards a little man from the front office. He leads the way back to his booth, locks up, and gestures towards his moped, climbing on. I climb on behind him, wincing. It's like sitting on a griddle. We lurch into the cemetery. After about half a mile he pulls up at a junction, pointing at a sign carved into a post.

"Protestant Cemetery".

Ahhh, right. Of course. I dismount. He waves his arm in an expansive gesture, towards the slope beyond it. He revs up again, pulls the bike round, and scoots off into the dappled depths, slightly less sombre than before. I'm on my own again, in the forest of graves.

I revert to systematic mode, weaving up and along the rows, now finding most of the names in English, some family plots in Anglo-Greek. Then I can see the shed, in the corner, up ahead, and from that I can orient. Around it lie barrows, bins, tools, bags of whitewash and cement, broken slabs, pipes. The maintenance yard. The edgeland of the graveyard, the functional bit that helps keep the rest in some kind of order. A work ethic prevails here, a different mood. Tim might have liked that. He was a potterer, a man who tried to learn practical skills. And there, ahead of me, is his slab. Was I close yesterday? Not even.

I sit down beside the tomb for some minutes, for a think, and to reflect. I wonder how many others might have done likewise here, before me. The grave is dark, gathering primitive plant growth, obscuring the words, and the sword. A cypress casts a relieving shadow over him. There are no embellishments. The photo had a small flower urn – empty – but that is long gone. I place a dove's feather on the slab, by way of a humble offering. The inscription reads:

T.H. WHITE
1906–1964
AUTHOR
WHO
FROM A TROUBLED HEART
DELIGHTED OTHERS
LOVING AND PRAISING
THIS LIFE

Below this a sword is engraved on the stone.

* * *

Why have I been looking for him? I think it's to help others re-find him. If England cannot have his bones, perhaps in lieu of those it could have a simple memorial. Perhaps with the coming 50th anniversary of his death, January 2014, something might be arranged.

I cast my mind back to the note I received from his old school when I made a written enquiry to them about the whereabouts of his cottage. It was a short, polite reply, a while later. They didn't know about the cottage, but offered some helpful suggestions, and this: "you may be aware that he had to leave the school in something of a hurry". He doesn't feature in the school's lists of celebrated alumni, while actors, journalists and businessmen do.

While teaching, Tim White had been trying to make some money by writing what might be termed 'racy' novels under an assumed name – James Aston. This was discovered by the pupils when the publisher's letters were mistakenly delivered to a pupil with a similar name. While White's parting from the institution may have been swift, it also seems to have been amicable. It's clear from some of his published letters that he remained on friendly terms with the headmaster, Roxburgh.

Nearby, the Hoopoe is digging in soft earth in the lee of the boundary wall, probing with its sickle beak, crest rising and falling. Loaded up, it wafts on fanned and shut, fanned and shut wings up to a cavity in the wall, and feeds its clamouring offspring. Perhaps it would have led me here, if I'd persevered on that notion.

I think I'm working out why someone like Tim White would attempt to find a wild and free state of being by getting a Goshawk. I'm sure he wanted to tame the bird, wished it to trust him. But I suspect that, subconsciously, he may have been willing it to escape.

I turn once more to the grave. The feather is gone.

Friday

A young, recently fledged female Sparrowhawk has been visiting the front garden for the last ten days or so. Today it flies up and out of sight from below the bedroom window. It then returns, and flies in a rather untypically ragged glide into the wind, and lands in the willow about 50 metres away, to sit in the midst of the tree. A Woodpigeon has been watching it all the while from a small Ash tree in the garden. After a few minutes, I notice the Woodpigeon fly towards the willow and flush the hawk, which then departs, beating off low. I have no doubt that the Woodpigeon has 'seen it off'. One for the experts again.

"Can Woodpigeons in a sense 'stand up' to Sparrowhawks as long as they 'can see them'?"

It's new to Mick. "I have not noticed the like of this before," he says. "Sparrowhawks are eating Woodpigeon nestlings at the moment – generally that continues through to October. Presumably an adult Woodpigeon might well chase a Sparrowhawk off from the vicinity of its nest if the circumstances were right. I have seen a Mistle Thrush charge at a male Sparrowhawk once, and chase it off. Also, I once watched a male Red Grouse fly at and strike a yearling female Sparrowhawk that was trying to grab something from under the heather."

SEPTEMBER

I'm cycling to work, and a Buzzard-sized raptor flies out from the side of the bridleway about 30 metres ahead of me. It is visible briefly, then flies across and behind the hedge on the opposite side. What is immediately most distinctive is a luminously white band around its rump area. I look for prey remains and find none, then I turn and the raptor is gliding over again, higher now, perhaps to have a look at me. Again the white band is clear.

Unlike a harrier, it isn't holding its wings higher than level. Nor does it have a harrier's buoyancy. It causes a stir among the corvids, and I note too that the Pheasants have just been let out and are all over the place (in every sense). It's a place that harriers do pass through, but I'm just wondering if it might have been Gos.

I describe it later to Mick. "I know this white band might suggest ringtail (female or young) harrier, were it a rump, but I'm wondering if a Gos side-on might show such a white band – the under tail giving an impression of being on upper body too? Or is this just in spring display?"

"Yes, the white under-tail coverts often curve up the side of the base of the tail, and sometimes on juveniles, the largest upper tail coverts can have pale buff tips," he replies. "However, Gos never really look as though they have a white rump. Harriers don't get such a reaction from crows."

He has another suggestion. "What about Harris Hawk? They fly differently and have a very dark tail. Otherwise, not sure if I can help that much."

"I realise it's not much to go on. The startling number of poults running around all of a sudden across quite a big area would be a real magnet to any Gos passing within a few miles of here, and I guess harrier too. Harris Hawk is a good shout ..."

Mick provides a quick update: "Dispersed juvenile birds on the wing up here now with sightings occasionally in dry spells after prolonged rain. Adults up in the air a wee bit but not much. Searching woodland blocks for body feathers and getting the odd flight feather. At the end of the season the number of young was down a fair bit, but it's enough to keep things going next year."

Mick also reports finding something interesting in the woods. 'The recently stripped bones of a Mountain Hare, together with a streak of shit and seven body feathers of a two-year old Goshawk. I presume the bird is a female, that it took effort to subdue the hare, and time to consume it – losing feathers in the process.'

"Interesting crime scene! Adult hare? 6 kilos vs 1.5, or thereabouts? Quite a struggle I'll bet, and noisy."

"Probably not as mismatched as that. Mountain hare adults are 2.5 to 3kg. This juvenile would have probably been less than 2kg. That said, it is still a big item, and would have delivered a hefty kick had the hawk fumbled it. If it was the same two-year old female that bred nearby, it was a bigger than average bird – quite possibly 1.4kg as you say. I hadn't thought of the noise – yes hares squeal a fair bit when grabbed.

"The hare's hind feet were 137mm, and bone ends not completely ossified, so it was a juvenile. Here in north-east Scotland mountain hares are taken often by Goshawks, but rarely adults; most are leverets recorded from May onwards, with the biggest taken in June and July when the female is hunting. They usually leave the hind legs intact, so I have lots of measurements – it could be a paper one day."

Meanwhile, in my Gos-free world, I muster some prey remains of my own: "I found a Moorhen leg today – outside the Tate Modern. I was at the Peregrine viewing point on the South Bank, on a story for *Lost in London* magazine. There are around 20 pairs of Peregrines in Greater London now."

Sunday

There's bad news from Malcolm in the Borders. "I got a new nest this year near [town in Lothian]. On my first visit on 26th June there were two chicks in the nest. I went back two days later and both were dead, shot with shotgun."

I ask him how the few Golden Eagles are faring in the Borders since the poisoning of the young bird a few years ago. More bad news. "It's a disgrace since the poisoned bird. At that time we had three pairs in south-east Scotland. Now we are down to a single bird."

There's more disturbing coverage in the media. Further details have now come to light of the 'gruesome' incident in Devon that my colleague had alluded to in the spring. No fewer than four Goshawks, and a Buzzard, were found in Forestry Commission woodland near Exeter. Tests have shown that all were poisoned by a cocktail of pesticides. Rewards have been offered in return for any information on the perpetrators.

Gamekeepers' representatives are saying encouraging things in *The Times*. Other gamekeepers have come forward to report illegal practices on their estates – courageous acts on their part. It has prompted the launch of a 'Good men stand up' campaign, which offers support and advice to other keepers who might need and wish to follow suit.

I've had an update from Paul Marten in Sussex. Good news. "The pair I watch had another fantastic year, successfully fledging three young, a male and two females, as far as I could tell, judging by their sizes, especially their feet. They moved to a new nest, some 100 metres from the old one, and the cover was almost non-existent, so, rather than risk disturbing them unnecessarily, I just popped in now and then to check everything was ok. Obviously, rearing three was proof that all was fine. That's seven in two years now ... bloody brilliant! I've noticed that when I've been watching them perching near the nest, usually just as they fledge, they try and look skinny, much as a Long-eared Owl will do, and hope that I can't notice them."

Monday

A timely paper on historical vermin records has appeared in the journal *Scottish Birds*, published by the Scottish Ornithologists' Club. I've borrowed Mum's copy. Author R. L. McMillan has had access to an almost continuous list of what was killed between 1867 and 1988 (after which time the killing of protected species was not recorded) on the private Atholl Estates in Perthshire, the second largest in Scotland, close to the place where the last Goshawks to breed in Britain are reputed to have lived, and been killed.

It certainly supports the idea that astonishing numbers of raptors were being killed. Between 1867 and 1911, well over 20,000 hawks, owls and Ravens were taken. The term 'hawks' covered any day-flying bird of prey. An average of 260 were killed each year, supporting the idea of a constant influx of replacement birds: the sump effect, again. There is no suggestion that keepers would exaggerate their vermin bags as no bounty system was in place. If anything, the numbers might have been under-estimated.

The estate was rearing and shooting grouse of all kinds, and Pheasant rearing was increased from the early 1890s. Around this time there is a noticeable further increase in the number of hawks, owls and crows killed, which may reflect additional control measures put in place around Pheasant pens. Pole traps would have been deployed – leg traps placed on top of prominent posts – although these were soon deemed barbarous, even by the standards of the age, and outlawed in 1904. By this time there were none bar the occasional passage Goshawk from the continent left to protect.

Not all gamekeepers were prepared to kill protected species. At the start of the 1960s, one keeper spared hawks, owls and Ravens. Others followed suit, believing it unnecessary to kill these species, even before the law changed in 1954. It is clear that some keepers had the integrity to do this, and that the Estate as a whole was prepared to grant them this latitude. It is also clear though that strengthened wildlife protection laws have been little impediment to those prepared to flout them. The 1954 Protection of Birds Act made little difference to the numbers and species killed by any indifferent keeper at Atholl.

In the years before they stopped writing down what was being done illegally, "The head keeper's beat on this single large estate, between 1980 and 1988, accounted for nearly twice as much illegal raptor and owl mortality as was officially recorded in the whole of Scotland." The RSPB has often stated that the recorded incidents of illegal raptor killing are merely the tip of the iceberg of activities that by definition go mostly unseen, and unseeable. Here is further supporting evidence.

OCTOBER

A Harris Hawk has been seen near my village in the last couple of days, selling local birdwatchers a dummy. They've been out looking for a reported Pallid Harrier straying in from the continent, and found this bird that has strayed from a falconer. This may explain the odd bird I saw a month ago, when I was cycling to work.

For most of the autumn I have been tucked away writing *Silent Spring Revisited*, which leaves time for only occasional forays to look for Goshawks. I do receive this report from my friend Charlie:

"I went looking for the reported Pied Flycatcher earlier and when walking towards the old heath heard this quite chilling, screeching raptor, which I didn't recognise but assumed a Sparrowhawk. Anyways, just checked on the RSPB website. What I heard sounds much more like a Goshawk – and no, definitely not a Green Woodpecker! Possibly one on steroids. Anyway no sign of anything, but fairly sure that's what I heard. Will listen to it a few more times before reporting it though ... Charlie B."

Other than this, all Goshawk leads locally have gone cold, and I am beginning to wonder if I have a story to tell at all. There is little more to report until mid-December, when an email arrives, out of the blue, from a man who lives in the fens. Subject: "Goshawk".

"There have been several sightings this summer and autumn of a Goshawk in the village around the area of the church and the adjacent old parkland which has some mature tree cover. I wonder if you are aware of this?"

It being the fens I have to admit I am sceptical, but prepared to suspend my disbelief. By now I will grasp at any lead going.

"Many thanks for getting in touch. I hadn't heard about this. I know that area quite well, so would be interested to know more and maybe take a look the next time I am passing. If it was present in summer it suggests more than just a passage bird."

I get a little more detail. "Observers are suggesting to me this is a female Goshawk and may well have a mate and could even have been feeding young. I could put you in touch with a local who has been observing these birds quite regularly. Let us know what you would like to do."

The plot is thickening. This will have to be followed up. "Yes, I would like to meet the person concerned. As a matter of interest, how did you know of my interest in Goshawks?"

"I googled 'Goshawks Cambridgeshire'," he explains, "and arrived at a discussion over a Goshawk sighting in south Cambs which you had been involved in. I think we may also have met somewhere in the past. I used to work at the National Trust. Will be in touch when I next meet the Goshawk observer, who incidentally is a very competent ornithologist."

There would be an opportunity at the new year. "I should be passing at the end of December, en route to Norfolk, and early January, heading back," I tell him. "I'll take a look – with low expectations of seeing anything! These birds can be quite extraordinarily secretive."

"Make for the church and go north down the lane some hundred metres," he advises. "The Goshawk sightings have been either side of the road in trees behind gardens and in parkland opposite. Have a good Christmas."

I check on recent discussion of Goshawks in the county. There's a report from earlier in the year, under the heading "Another one that got away", by someone called David.

"While doing the school run this morning I caught sight of a raptor that was flying low over the fields... Something about this bird made me take a second look. I have seen many Sparrowhawks recently, however this bird appeared more bulbous, 'full-chested', with a more laboured flight on rounded wings. By the time I had stopped the car the bird headed off north. I had no equipment with me which is typical, but I think I may have had a Goshawk! Have there been any other sightings in this area recently or do they breed nearby?"

Local recorder Mark Hawkes confirms that there have been "no confirmed breeding records in the county in many a year. Still a rare bird in the county, with perhaps two to four records per year. This species suffers from lack of submissions to the Records Committee, making a clear picture of its true status in the county not really possible – my guess would be that they are out there and breeding somewhere in Cambs."

Friday

I have a rake at the fen town just before the year ends. Looking at the map later I realise I walked down a different road to the one described, although I did have a quick look round the church itself and the road in question. There is a rookery there, which might attract a Goshawk. But if there is Goshawk – or even Goshawks – there, I don't see them, and don't find any signs. To be honest, I never expected to, although these witness testimonies do sound credible. I would be keen to know more from the main witness herself, as and when.

A list has been compiled of all the species of everything seen in the past calendar at or from The Lodge. The British Trust for Ornithology, based at Thetford, have recorded more birds, while The Lodge recorded more species of all kinds. It's all good fun, and at least they got Goshawk. It hasn't escaped my notice that not a single Goshawk has been officially recorded locally in the whole year. I am fairly sure they've been here, but I wonder if now they have gone.

Chapter 10

FENLAND

JANUARY

At Derby Crown Court, a gamekeeper is appealing against his conviction last June on seven charges relating to the illegal use of a trap baited with a live pigeon in order to take birds of prey, the intentional taking of a Sparrowhawk, and a number of other animal welfare offences. At the original trial last year he had been sentenced to 100 hours of community service, and to pay £10,000 in costs. His defence team have launched a 'blistering attack' on the integrity of the RSPB, claiming that evidence has been planted. The hearing will run for some time.

Tuesday

I have spent most of the mid-winter break tucked away, writing. I can now turn my mind to properly investigating these improbable yet oddly persuasive reports from fenland.

Jon, who first alerted me to the possible fen Gos, has returned from his winter holiday. "Just met up with Jane while walking dog at the fen. She confirmed the sighting of Goshawk in her garden before Christmas. She would be pleased to hear from you."

I get in touch right away. "I'd be interested to know more," I tell her. "If I'm passing that way I'll let you know in case there's a chance of a guided recce."

At the same time I get a message from Daniel, a friend who works as a barrister in Cambridge. It's headed "Lunch on Green Street".

"We witnessed this scene yesterday lunchtime. I assume it's a female Sparrowhawk."

I'm trying not to get too far ahead of myself as I fumble to open these attached photos – Cambridge isn't far from the fenland site with the putative Goshawk. Could it be that it or another Gos has been hunting in the city? Another possibility is Peregrine, which I know has adopted Cambridge recently. I've seen one over the market square. The photos reveal a sequence of images of a hawk which has caught a Feral Pigeon. It's not a Gos, or a Peregrine. It's a Sparrowhawk. I'm a little disappointed, but it's still noteworthy. I've seen numerous photos by now of Sparrowhawks in action like this, but this is the first truly 'high street' incident of its kind that I've been made aware of.

"A crowd of bystanders watched in quiet amazement," Daniel writes in his accompanying notes. "Although a few just walked or cycled by apparently oblivious. One or two people were getting just a few inches from the bird to take photos and, although it watched them, it showed no fear. At one point a woman tried to frighten it off, saying 'go on, shoo!' The hawk looked at her – I would like to think with some disdain – and then flew, about a foot off the ground, 20 feet down the road with its prey.

"It was extraordinary that the bird conducted its business in front of so many people, and I wondered if it was either very hungry or used to people. I observed the scene initially for about 15 minutes, not least to make sure the hawk was going to be OK, then went to meet my friend nearby. We came back to watch for another 10 minutes. I returned after lunch to find three piles of pigeon feathers and just the pigeon's head, eye still open, remaining. A memorable lunch!"

I seek an expert German perspective from Rainer: "This is an unusual occurrence here. Do you have Sparrowhawks this 'brave' in German cities?" I ask him. "If not an escaped bird, it could be a particularly hungry and desperate bird, even a passage bird?"

"Yes," is the verdict. "We have, not very common, but regularly. And I do not believe that these birds are particularly hungry or desperate, neither do I believe that they are migrants. On the contrary, there is good evidence that most of these birds are well fed, and that they are simply well accustomed to men and to an urban environment. The bird seems to me perfectly healthy and alert. You remember our young female Goshawk from Berlin, plucking a crow just above our heads? Also this Goshawk was far away from being particularly hungry and in no way desperate."

I've also sent it on to Professor Newton. He confirms that it's an adult female, and maybe an unprecedented occurrence, in such a place. "I wonder if it is a released falconer's bird. It has no jesses, so we have no way of knowing, but the same thing could happen again there," is the world authority view.

Thoughts are returning to the fens and, in line with Prof Newton's deductions, I wonder whether a released or escaped male Gos might be at the root of these reports. Then again, if the urban situation is actually the best available habitat in a fenland context, then perhaps a Gos could already be learning to cope with the proximity of people there. If the Gos can be urbanised like this on the continent, why not Sparrowhawks here? But then why so little evidence of it up to now? The puzzle preoccupies me for some days.

Tuesday

Breath has been held in anticipation of the verdict being announced today at Derby Crown Court. Judge Watson has dismissed the appeal, and the gamekeeper has been ordered to pay another £7,000 in costs. "Industry leaders and employers need to do more to stamp out these crimes," declares RSPB Conservation Director Martin Harper after the case closed. "We believe that land managers and owners should be held legally accountable for any wildlife crimes that are committed by their staff, as is the case in Scotland."

Mick had been called as an expert witness, as he was for the original court case: "I am not long back from Derby," he tells me, "initially a bit bruised and pessimistic from cross-questioning by the defence – but then elated at the outcome. Your colleagues worked hard for this result – perhaps we might now see an increase in Goshawk in the area. Much depends on how other keepers respond – they must be feeling less secure about the protection that might be expected of their employers."

There have now been more than a hundred gamekeeper convictions for offences against birds of prey since 1990, and it needs to be remembered that these convictions are extremely difficult to secure.

FEBRUARY

I've received some more detail from Jane, out there in the fens. It seems the Goshawks she thinks she's been seeing might be more than just winter visitors. The time of day of these observations is interesting too. She's an early riser, just like the Goshawk. "It would have been in mid-May when at about 4.40am on several mornings I heard a single raptor cry, which was almost familiar but I knew was not a Sparrowhawk. I didn't hear it at any other time except first thing, and one morning I heard it and immediately went on the RSPB bird identification site to listen, and concluded it wasn't a Sparrowhawk but sounded very like a Goshawk. I have seen them in Yorkshire and in Austria, and am aware here is not typical country, but a couple of summers ago I was not far from here when twice in a period of about three weeks and when out with a more experienced birder we had seen a bird he identified as a Goshawk.

"I am not easily persuaded and like to see a bird before I confirm it to myself. In late May/early June last year I began to hear cries throughout the day in two locations, about 500 metres apart, one at home in the copse opposite and the other on the fen between. The copse is difficult to access, so early one morning before the rest of the dog walkers were up I went down to the other site in search of the calling bird. I crept along the tree line and saw a male Goshawk in a tree. He took off. Their flight is different, the white eyebrow and size and colour of the eye – no doubt in my mind it was a male Goshawk. A couple of days later in the same location I saw a large, brown raptor which was not a Hen Harrier (there is a female about on the fen) or a Buzzard.

"I work on the Sherlock Holmes theory: you work through the evidence and whatever you are left with – however implausible – is the correct answer. Someone else in the village has also seen it but I have not had the opportunity to speak to him about it. A fellow dog walker has also seen a Goshawk in recent years.

"About two weeks before Christmas my neighbour rang me to ask what was standing in her back garden consuming a Woodpigeon. I identified it as the male Goshawk which, when it flew off, effortlessly lifted off the ground with the prey and flew within 15 feet of the window from where I was observing – white eye-stripe very apparent, and also its size and thick legs. Would be happy to show you where – several locations but in quite a small area including my neighbour's garden!

"If I can get a photo I will. I'd be pleased to meet with you and in the meantime will keep you appraised of further sightings. Just to give you a bit of background I have been an RSPB volunteer for about ten years."

I reply immediately. "It could be dispersing juvenile birds, maybe even attempting to set up territory. Hard to be sure. I'd be very interested to stop by and have another look, so will let you know when that might be possible."

I wonder if they have got as far as nest building, but there wouldn't be much peace and quiet for them around there. It seems there is a body of evidence rather than any single conclusive piece. I will head up there to walk through it with Jane. It's raised ground on fenland, and about as treeless as it gets, so it shouldn't be difficult to find a nest, if there is one. If these reports are right, it could be dispersing juveniles. I seek the opinion of Mick.

"Sounds fairly convincing," he says. I'm pleased he thinks so too. "Birds often call in several places so it's not necessarily two birds. Either place could end up as a breeding site. If birds are new into the area, an adult male with juvenile female is expected, but not so good if the male has been there for a few years and keeps getting a young female. Either way, provided the lady is correct, an adult male living in an area and calling is great – he will get a female onto eggs eventually. Best of luck with it."

"Thanks Mick. I guess I'm looking for a nest, especially if access isn't great for a thorough scour? I can't overstate the lack of woods locally. Have a look at this map: the location of Jane's sighting is indicated by the marker. There is an active rookery in a cluster of trees there, and a belt of trees not far away. The fens would be a ready source of prey of course but it's well covered by birders and there are very seldom reports of Goshawk. Anyway, all adds to the intrigue."

Mick may have underestimated the treelessness of these southern fens. "I see what you mean – distinct lack of trees," he says. "I am not optimistic but would love to be proven wrong. Certainly makes the job easy – just check the thickest bit of tall cover. If there is a rookery, there will be prey remains."

A chance to visit comes up. I contact Jane. "Would you be around late morning tomorrow, when I'm passing through? We could meet at the church, for example – weather permitting."

"I'll meet you at the church south-side door," she replies.

Saturday

I'd hoped to get there in the morning, but the fog that settled on another freezing night still envelopes the landscape, and shows no sign of lifting. But by lunchtime the sun has begun to bore a hole in it and the day to brighten, and in due course the landscape is shimmering again under clear skies. I head over to Cambridge, teetering in an ice-rink car park with a 20-kilo sack of birdseed along the way, then to Emmaus for a cuppa and to browse their second-hand books. I find a Folio edition of a rare T H White book, *The Age of Scandal*.

So on this bright and bitterly cold morning, with temperatures plunging to minus 10, I meet Jane at the door of the church that dominates this small area of elevated land among the stretching fens around it. We walk on the churchyard's ice-caked path to the road to begin the tour. I ask her to start from the beginning of her various Goshawk sightings. There have been more, even since her last email of just a few days earlier.

I get the full story as we walk round her neighbourhood. She is a distinctly credible witness, and it really does begin to sound like there are Goshawks here. At face value, a young female – like the bird in the glass case, the bird in the Berlin park – has been hunting around their gardens in recent days and weeks. Jane is familiar with Sparrow-hawks, which she also sees. I think we can rule them out as suspects. "June to August last year on the fen and near the house I saw a male and also a juvenile, brown with teardrop vertical streaks." Jane's description of the hawk's size and striped breast, and that she has seen it going after Rooks, clinch it for me. Her reports of the bird calling at dawn from a dense spinney, with the rookery in some of its taller trees, throughout May and into June is fascinating – could this have been an unpaired male? Her description of the male that she got close enough to for eye contact intrigues me – can a wild bird be this tame already? Or might this be an escaped bird? Is it possible that there's an habituated male, escaped, hanging around, and attracting the young female in? Jane was able to approach it at dawn as she walked her pet lurcher. She describes how the hawk 'glared' at her then flew off. Even a wild Sparrowhawk wouldn't usually be that approachable, I think.

I can see no sign of any nest or even anywhere obviously suitable, but that could be further afield. Jane promises that if she hears any more of the calling this spring she will let me know and I can head back there. I will also be able to have another rake next Sunday.

Mick thinks it possible a wild male Goshawk could be confiding in certain circumstances like these. "Sounds pretty good to me," he urges, when I update him. "Could be that the male is just habituated to folk – if there is limited woodland cover it might just stick with what is there: a bit like urban hawks that habituate because there's no risk and, with lots of folk about, no other choice."

I hadn't expected this to happen for some years yet, if it were ever allowed to happen at all here.

"Looking forward to the next chapter in this story," he tells me.

I know where I'd like this to lead, but I'm frankly no more certain than Mick of where this plot is heading.

MARCH

I've heard no more from Jane for about a month, but a message has just come through. "I'm sorry but I haven't been too well since we met and life has got away from me a bit," she reports.

"My neighbour advises me that there has been a kill in her garden this morning, evidenced by feathers seen from her kitchen window. She hasn't investigated, but I will when she returns tomorrow and will let you know if there is anything significant for you to see."

"Really sorry to hear you've been unwell," I reply. "Hope you are making a full recovery. Remember to look out for displaying birds – this is the time of year. Sunny mornings best, but not exclusively."

Jane provides me with a summary of the Goshawk observations made over the winter:

• November/December whilst at my neighbour's saw large brown raptor fly into a field maple and the tree 'exploded' as the resident bird population dispersed very quickly.

• 9th February, while sitting in my neighbour's living room, I saw a juvenile with distinctive teardrop streaks fly towards the house and us: a large, brown raptor, effortless, swift. Saw it again about 15 minutes later a little further away as the local population of Woodpigeons and corvids rose as one from the ash trees in the vicinity.

• In the last couple of weeks I have seen on several occasions a large heavy raptor early in the morning, shortly after dawn, setting up Woodpigeons in the area across the road from my house.

Wednesday

I wake early on the first day of spring and push the curtain aside. A Wren is bouncing from the Ash tree to the lamppost, yelling. The Mistle Thrushes are ferrying worms from the muddy field entrance to their young in the iron rafters of the open-sided barn. My first Chiffchaff of spring is calling. I catch a train to Peterborough for a meeting of the *British Birds* board, noting newspaper headlines about what has now become the worst drought in England for 30 years.

Another message comes in from Jane, subject: "Fresh kill".

"My neighbour called me at 8.30am to advise that a 'large, brownish raptor, lighter underneath with streaks and thick legs' was plucking a Collared Dove about 30 feet from the house. Unfortunately I didn't see it but when I asked was it bigger than a Woodpigeon she said 'Oh yes, and chunky, stocky, standing upright'. This is the third kill of a Collared Dove in the last couple of weeks in her garden. Today there are wing and breast feathers, quite a lot of blood and a small piece of bony material that I have recovered (now in the freezer). Meant to ask, do Goshawks stoop as Peregrines do? I have seen a large raptor stooping – not a great view as into the sun."

I reply that yes, they do stoop at prey, although it's rarely if ever mentioned in bird books. I promise to get there at the weekend.

Monday

I return to the fenland town, park up and walk in the warm spring sunshine to an isolated oak in the middle of a field on the edge of town, select
a comfortable tree root to perch on and sit down to watch. The air is full of
birdsong and bird activity, especially Rooks organising Rook society, and
Woodpigeons commuting between town and country, with much else besides. The oak is not yet in leaf so my view of the sky is mostly unobstructed.
I have been *in situ* for less than ten minutes when a hawk goes up from the
crest of the gently sloping pasture in front of me. It circles quickly, rising, as
though winding itself up, and then goes into a long, smooth diagonal stoop at
the houses on the near edge of the town, ignoring the Rooks as it goes – they
have clearly seen it – and hitting a garden with an audible clatter of leaves
and pigeon wings as birds scatter in all directions from beyond the boundary
hedgerow.

Now, my first thought when I saw this hawk was that it was distinctly
brownish and fawn-coloured below, and that it was female Sparrowhawk
size. But the sheer velocity and power of the dive and attack made me think
again – was this the young male Goshawk?

I have no idea what the outcome was of the attack. So I sit tight. Before
long, I notice hawks in the air high above me. One in particular is owning
the place, gliding around, a Great White Shark to the Rooks' Sea Lions. I
watch this pair on and off for about an hour, they are mainly soaring, with
some roller-coaster display flights. I have the sense that both birds are young
ie quite tan-coloured and not noticeably pale below, which tallies with Jane's
description. There is great consternation among the resident Rook colonies,
and the Woodpigeons. The corvids don't seem to know what to do with
themselves, as the hawks circle.

I move to get a clearer view, unobstructed by the branches of the oak,
and lie back. The demeanour of the other birds certainly suggests Gos. One
hawk has a tilt at them, scattering them. In the open air, one Rook is bold

enough to approach the hawk, then scrambles comically back to the flock when the hawk turns the tables with a mere flick of its tail. The air is filled with palpable crow anxiety. The current birds both look sub-adult but I am getting closer views of the male than what I take to be the much higher female. The hawks come and go, on and off, for a good hour. After a gap, at one stage there are three of them really high overhead. One peels away from the pair and heads off.

I can't think of many ways I'd rather have spent my birthday. It has been a memorable, euphoric afternoon discovering that these reports of Goshawk from a small town in the fens are accurate.

Jane is really pleased when I tell her what I've seen, and I meet her neighbour too – the Vicar. "It's always good to know that I haven't been imagining things ..." she tells me, handing me the prey remains in a polythene bag. I ask her to let me know as soon as there are any further developments. "We'll keep you posted," she promises.

It feels like such a breakthrough, and so far ahead of schedule, in terms of urban colonisation. I'm minded to keep them quiet for now. I will return, to try to identify their focal point.

I get in touch with Mick to relay the good news. "It was amazing, and the absence of woods or many trees allowed unobstructed views for long periods. I couldn't locate a nest, although the options for this are very limited, and most of the big trees have Rook nests in them. The hawk came overhead and had a look at me a couple of times, and didn't seem too fazed. Jane has been seeing/hearing Goshawks since last May/June in this area – I would guess it was the male (or another male) then – possibly a pair. So they may have tried and failed already, or got there too late to try at all last year. Hope you are enjoying the heatwave – I heard you had your hottest ever March day yesterday!"

"This is great news," says Mick. "Not only colonisation, but of an area where roost sites/nest sites are clearly less than usually preferred. We get the odd situation like that here, but the birds wander off to adjacent woodland when they mature. Your birds are obliged to stick with what they've got. There's plenty of food, so we have to be optimistic about it.

"Birds can breed in their first year – common in females, rare in males – but it happens where there is lots of food and no adults ... Breeding attempts with first-year males tend to be late and often fail, but are usually followed by successful breeding the following year if the birds are still alive. My apologies, I am getting way ahead of the situation. Hope you get more on these birds before too long.

"Here, it is an early year for breeding. John Young telephoned in this morning to say that his earliest nest could have laid a week ahead of previous lay dates. I am still spending much time overlooking swathes of ground where game interests are high and the (few) Goshawks appear to be first-year birds. The odd eagle and harrier are also first-year birds – for hawks this terrain is the land of eternal youth. Looking forward to the next chapter from the fens."

Wednesday

Back at The Lodge. From the first floor office window I notice three raptors flying in file across the quarry from west to east, just above treetop height. I keep a pair of binoculars handy and I get them focused on the birds. I can quickly make out that the bird in front is a hawk. It looks medium-sized, from this side-on view. The two behind are Red Kites – an uncommon but increasingly reported sight here. What is most odd is that the Kites are clearly chasing the bird in front. The leading kite has a go at the hawk. I can't recall Kites ever looking quite so exercised by another bird. Would they be likely to be so perturbed by a mere Sparrowhawk? They disappear from view beyond the trees and the building. I quickly move outside in the direction they had been heading, notifying a colleague on the way.

A few minutes later two kites – for sure the same ones – are circling, gaining height over The Lodge gardens. There is a hawk with them. All three circle together, with no sign of any tension. The hawk now looks small along-side the gangly Kites. Because of what I saw minutes earlier I am still wonder-ing if it could be a male Goshawk. I never see it 'flutter' like a Sparrowhawk. It is soaring, with the occasional gentle flap – leisurely circles, not tight. It just doesn't look big enough, though.

I am now starting to doubt what I've seen in the fens, and having a crisis of confidence. I'm racked with self-doubt. There were lots of raptors soaring and moving around on Sunday, typically for the time of year and the ideal conditions. My friend Steve Dakin, Professor of Visual Psychophysics at University College London, once told me this, when I asked him about how we can ever be sure what we've seen. "Witness reliability has been measured and it is generally atrocious," he said: "Worryingly, witnesses tend to remain confident about dodgy recollections. Our vision – like our memory – is just 'a story our brain tells us' to make sense of our various experiences."

I am convinced something happens to our vision that distorts our im-pression of hawk size. And I know my own mind's eye doesn't retain detailed images – it stores impressions.

I shouldn't need this but I go back to Mick, and he reminds me of the basic differences between the species: "Gos is generally heavier in the body – bigger head and deeper chest than Sparrowhawk. Wings more pointed and tail shorter and less square. First-year Gos have a slightly longer tail. Gos in aerial display just now are predominantly young or unmated birds, but still some interactions on sites where females are competing for males."

I have to accept it was a Sparrowhawk circling over the gardens – but perhaps it wasn't the bird I'd seen the Kites chasing minutes earlier. Why would they bother with a Spar? And why weren't they mobbing it subsequently?

Without giving details of place I've discussed with one or two people the possibility of Goshawks in fen country. An ornithologist friend who lives up that way is extremely sceptical. "I did actually see a Goshawk once in fen-land," he told me. "It was causing mayhem with a flock of woodpigeons in a well-wooded place. But in the intervening 10 years I've not had a sniff." He makes me begin to doubt my own sanity. I become desperate for clinching proof. That's the one thing we don't have.

"Young Gos will adopt sub-optimal habitat," I tell him, probably unconvincingly, too keen to justify myself, retrieve some credibility. "Young males breed late and often fail, if they breed at all, in these places. There is an absolute ton of food for them up there in the fens. But let me check again."

***

I can't concentrate on or think about anything else. I have to know what I've seen. Have I been imagining things? Have I merely been seeing what I want to see, despite myself, despite everything, all the training and theory? I have to try to prove that I've seen Goshawks at the weekend.

I take the afternoon as leave to go back to the fen site; another hot day in a spring now desperate for rainfall. I park up in the quiet lane near the church. There are already Buzzards in the sky. I walk just a few paces, watching them, and I get chatting to a couple in their garden who have seen me watching the sky. They ask me what I am looking for. I am non-specific, and before long they tell me – unprompted – that they've seen a Goshawk ...

I take some encouragement from this, but the first hawks I find today are definitely Sparrowhawks – singles and a pair at different times. The fluttering flight and the sense of smaller size and lesser power are immediately plain to me. The wings are translucent in the sun. There is none of the *attitude* of Sunday's bird, none of the agitation among the Rooks and pigeons present. I begin to wonder if on Sunday there was just a juvenile male Goshawk, and not a female, and sometimes Sparrowhawks in the air too, high above. I can't honestly claim I saw a hawk on Sunday on the scale of a female Goshawk.

"Pity," says Mick, consolingly, when I tell him of my doubts. "It looked like a good situation. If the birds are still alive, they could be in the next nearest block of woodland, but I guess with restricted access for you that might not be very helpful."

As it happens, on the last day of the month, a systematic watch took place across the shire, to record all passing birds of prey. 'Hardy observers spent one or two hours counting raptors at 12 sites across the county. For 10 sites we can summarise as follows: 5 species were seen: 70 Buzzard, 25 Kestrel, 9 Sparrowhawk, 15 Red Kite and 5 Marsh Harrier."

No Goshawk.

APRIL

I try again on April Fools' Day. Perhaps fittingly – and I know this may sound a little eccentric – I've brought a Carrion Crow I found dead by the road on the way, thinking it might have some useful purpose in death, as a Goshawk magnet. I place it in a likely place in the field, positioning it upright (propped on sticks), and retreat to perch on the feet of the oak, to watch, and wait.

Birds sort themselves out in the breeding season. Some species nest in colonies, or loose colonies, or are reasonably tolerant of their kind nesting nearby, while others like an exclusion zone. Smaller birds organise this on a ground-level, intimate kind of scale. They use vocal signals – songs, we like to call them, somehow allowing this anthropomorphism to slide when so much else is frowned upon – as well as visual, and their spaces are fairly small. Raptors, on the other hand, sort out their territory issues way above us, almost entirely unnoticed, because we don't look up that often.

Lying back, looking up, it seems to me that their chains of communication might traverse entire nations, given the quality of their eyesight and hearing, and the vast draughty spaces they occupy up there. Like widely spaced vultures watching each other as they scan the landscape below, so that dozens can converge on one carcase from miles around, so hawks and their relatives get the message from each other about where the vacant or occupied territories are, on the days when they get up there and posture. I'm speculating. I don't think we have any real idea what it is like to own the sky and the land below in this way, our own fields of view and terms of reference are so narrow, so myopic by comparison. We look up at the sky so seldom. And of course we have to climb hills and mountains or get in an aircraft to gain even a hint of the planet as a raptor enjoys it.

These rare sunny, warm, thermally days in early spring are raptor carnival occasions, the festivals during which they move around, shuffle their packs, spot the gaps and fill them, a country-wide gavotte, ending with pairings made and order established. Perhaps in this context the young, fenland Goshawk or Goshawks moved on, or back, to vacant space: to a more typical, wooded place. Or perhaps they caught a thermal and went all the way to continental Europe.

Not much happens today. There is very little action. Just one raptor – a Sparrowhawk – passes overhead briefly. What had so recently seemed like a Goshawk hotspot is now starting to feel fanciful on my part. Sunday feels like a long time ago. I find my thoughts turning to Thetford Forest. The Gos may have gone back there. Perhaps so should I.

Thursday

Some days have passed. I get in touch with Jane again. 'Any more sightings?' I explain to her the doubts that have been creeping in.

"On reflection, I can't be sure there's a pair. It is possible that a young male Goshawk might move on in spring to more typical forested breeding habitat. I can't really see anywhere locally that looks like a place they could nest undisturbed.

"Young, inexperienced hawks often nest later and their lack of experience often results in failure, even if they have a more mature partner. If both partners are first-year birds, their chances of success are further reduced."

Despite this, Jane has continued to witness visits. "I saw what I believe to be a Goshawk on Sunday last about 9am stooping in pursuit of a group of Collared Doves. We have both seen the Sparrowhawk several times and there have been no more 'kill' sites in next door's garden. I will keep you posted."

Rob sends an update from the New Forest. "I have seen Goshawks several times this spring, but no great displays or anything. Have heard them a few times too – the loud, deep, resonant yak-yak-ak-ak-ak-ak is very distinctive, both from a perch and when in flight. Or, the latter could have been a bird on a perch in the wood beneath the flying bird I was watching."

From what Rob has observed it sounds as though many of the displays turn into attacks. How better to impress a mate, after all? The male seems to hunt most in the first couple of hours of daylight, provisioning the female. She spends a lot of time 'egging him on' to do this and to build the nest.

"These local Goshawks don't always hunt in the stealthy, under-the-canopy way by any means", Rob adds. "They often, as they did this morning, circle up several hundred feet above a wood and patrol determinedly and accelerate into a stoop, Peregrine-like, raising the tail (like a singing pipit) to lose lift then plummeting at great speed, often with one foot sticking out – Peregrines seem to stoop to strike direct (as I also saw one doing this morning!) while Goshawks go into a high speed chase after a pigeon, or just disappear into the trees."

Rob has seen and worked it out for himself. He doesn't need me to tell him about the hunting by stooping thing, which I think has been widely overlooked in the books because of the absence of Goshawks to watch. I tell him about the recent study of the young pair on the edge of Brussels that nested in a low fir and raised five young, often catching Feral Pigeons by stooping.

***

Saturday

On a quick visit to Mum's I pop out for the morning to help check on a couple of Peregrine breeding sites in the Borders. I even manage to pick out two sitting female birds while out with Malcolm. Nice to actually find the target, and confirm there's nothing wrong with my eyesight, at least. We don't sniff any Goshawks, though, whose females should also be on eggs of course, by now, tucked away in the forests.

A scientific paper has come out that makes a few headlines in the wider media. It reports that Goshawks tend to single out odd-coloured pigeons from large flocks. Christian Rutz of Oxford University has been studying Goshawks in Hamburg for some years. Most Feral Pigeons are grey-blue but many flocks have a few white birds. Hawks that master the selective attack strategy are the best breeders, it seems.

Tuesday

Back home I search through the maps on my shelf and take down the one covering Breckland. I realise that Thetford Forest is closer to this fenland town than I thought. This could be the source of dispersing young Goshawks. It is certainly the closest known site for them.

According to Mick, with Woodpigeons abundant nearby, Goshawk pairs could be two to three kilometres apart, even less around the edges. This could give a potential population of 20 pairs at Thetford if there are no other constraints. Persecution could persistently take out birds from the edges of the big forest wherever there is good hunting (Pheasant release pens and intensive game shooting areas) so that the only long-lived and productive birds will be those in the centre – more than two kilometres away from game areas. Under such circumstances the forest might only hold three pairs; or none if all birds prefer to settle on the edge of the forest next to good foraging sites. I would hope this is not the case.

Mick is right. I find out that last year, only three Goshawk nests produced young at Thetford, of the six pairs present initially. One of the other nests collapsed under its own accumulated weight after some years of use and replenishment. In another place an adult female was found dead, apparently predated, presumably by another Goshawk, in the absence of anything else capable of the feat. We can't rule out that she was put there, dead.

I make some enquiries locally about the situation for Thetford's Goshawks. The Forest is owned and managed by the Forestry Commission, a public body. They have a duty to look after wildlife as well as the tree crop, and here, as elsewhere in the nation's woodland estate, they monitor Goshawks and manage felling operations and public access to minimise disturbance to the birds. There are places well known to bird enthusiasts where displaying Goshawks can be looked for in early spring. The best-known viewing site seems to have been abandoned by the birds recently, and another area has taken its place. While they remain so peculiarly scarce the birds are sensitive to disturbance and vulnerable as a local population to egg theft. Thetford Forest covers about a third the area of Kielder. Three successful nests, and one fewer pair than the year before. It strikes me as a peculiarly low tally. Kielder, for comparison, has ten times as many pairs.

There is no Raptor Study Group in Eastern England, but the Goshawk nests here are monitored by a local team operating under Thetford Forest Ringing Group. The group includes former Forestry Commission staff, so the birds are under close supervision. But to my knowledge there is no ringing of the young, and virtually no records of birds recovered in the surrounding area. So where do they go? Perhaps one day a tracking study can be done of the few young birds that Thetford produces, to try to answer this question.

I consult with Mick on the dispersal conundrum. "Yes, I share your thoughts," he says. "Perhaps hawks are out on the fens during winter feeding on pigeons, and come spring they return to larger woodland blocks to seek out breeding sites."

Saturday

I've come across a lengthy debate on the Internet about Cambridgeshire's Goshawks. Someone is looking for advice on where he might see them, now that the traditional site is no longer being used by the hawks.

"The only Goshawk I've managed to see was while watching a falconry display," says one contributor. "The hawk appeared in the distance and flew high over. The guy doing the display was convinced it was a wild Goshawk."

"Only one pair has been proven to breed successfully in the county in the last 10 years or so," says another. "And this is in part a result of disturbance at breeding sites. This disturbance is not always wilful and can be as a result of forestry work, recreational activities, bird-watchers, etc."

"I still don't think that it's a good idea in any way to put out information on the Internet that may in any way assist egg collectors or hunters or morons. These 'traditional' sites were reliable for certain species but this is, in some instances, no longer the case due to disturbance."

"No gen should be given on Goshawk locations," comes another view. "I've assisted the BTO in finding two nests in the last five years and the information is not going to be made public. Stick to the traditional site." Which would be fine if the traditional site was still active.

"Not the best thing to be posting over the webs guys!' interjects another.

"Point taken chaps! Suitably chastened. Mind you, my post as presented covered an area of 12.5 square miles ..."

"Moderators – can you delete this thread please?"

"Now you're being ridiculous. All I asked was if one fairly well known display site was still active."

"Yes but unfortunately in creating a thread like this you encourage people to 'let slip' about other sites and sightings, whether intentional or not, so not the wisest course of action. Anyway, let's hope requests like this don't become a habit."

"Not sure what the answer is here. Typically people 'in the know' don't keep info to themselves but include friends who in turn tell their friends and – hey presto! – a clique. My experience is that cliques like to bait others with odd comments proving, 'I know, you don't, and I can't tell you'."

"Interestingly, directions for [a new viewing site] are all over the web – including on RSPB local group sites."

"I was thinking of taking my mum to this, and then I read this thread. We never ever want to disturb birds, but how can the RSPB be encouraging it and then I read on here we possibly shouldn't go? I'm a little confused."

"The birds are seen high above the trees at quite a distance (probably secure over military land). There is no disturbance."

Where places like Thetford are concerned, many bird enthusiasts seem to 'accept' this situation of restricted Goshawks. They appear to blame eggers and each other's thoughtlessness or over-eagerness to see our few surviving Goshawks for the need for secrecy rather than the likely fate befalling fledged young. People really don't seem to have grasped this.

"We had all these problems here in the early years when two of the five nest sites were well known," says Mick, of the north-east Scotland situation. "There is a tendency for folk to keep breeding sites secret because they can suffer from robbing by eggers and hawkers and disturbance by photographers and casual birders. In retrospect, I now know that the poor population performance was because both fledged juveniles and adults were being killed. The biggest limit on Goshawk populations is still birds being killed away from the nest."

* * *

With the birds still so scarce here in the south, and in such high demand for glimpses, perhaps the tensions, jealousies and uncertainties are inevitable. Where this county at least is concerned, the Goshawk definitely isn't out of the woods.

Tuesday

As though to confirm this, news breaks of the latest destruction of a Goshawk nest in the Peak District National Park. It leaves just one active nest in the whole Derwent Valley. There were six at one time.

"Anyway," says Mick, "promising to be a good year here – over 50 nests we hope, and if this rain stops, maybe over 100 young. Aye optimistic! If you feel overly deprived and need time among the hawks, come north."

JUNE

Reports of Goshawk locally have dwindled. It hasn't escaped my notice either that Jane's reports have dried up. She's stopped encountering the birds in her town. Might this mean that something has happened to the birds? I am running out of places to look with any serious degree of expectation. I seize on any hint of a clue of one. A colleague reports a Hen Harrier over the pines on Galley Hill, close to work. Odd place for a harrier, I quiz him. What was it doing? Being chased by crows, he says, calmly. I cast my mind back to the Berlin female Gos, batting stiff-winged around that swing park, harrier-like in her exaggerated display posturing.

As for the rumoured birds near my village, my guess now is that they maybe tried to nest two years ago, as per the rumour, and that by last year they were more or less gone. There are consolations. I discover something else, when I'm not even looking. In fact I was barely even awake. I'm glad that others can vouch for what follows otherwise I wouldn't expect anyone to believe it. I'd even have doubts myself. I found a Golden Oriole. Now, that would be unexpected enough, but consider this: it was here in my Bedfordshire village, it was drizzling steadily (Jubilee Sunday – ask the Queen and Prince Phillip, who caught a cold), it was very early, and I was in my bed.

The bird was calling from the willow opposite the house. Was I half-asleep, in the optimistic inter-phase between dreams and reality (at such moments Father Ted's helpful diagram springs to mind), and merely thinking wishfully? Was this just a particularly melodic Blackbird having me on?

No. Golden Oriole for sure. I know this bird. It has haunted me since a boy, discovering them around French campsites. The first I ever saw was a pair that came torpedoing out of the forest canopy after a bird of prey gliding over the Ardeche river, making a tremendous din, worse than Jays, not at all what you expect from the demure image of the birds in the book. Their target may even have been a Goshawk.

Golden Oriole is one of those birds, the glam one that jumps out of the ID guides at you, the avian El Dorado. When you find the Golden Oriole you discover that it seems to be whistling its own name, in the musical voice of a Clanger in a bucket, and that this siren sound carries far on a breeze, and into your soul, from a source that is near impossible to pick out of the shimmering canopy leaves of the oaks and poplars wherein it lurks.

My Jubilee Oriole soon had bird lovers descending from near and far, forming an infantry unit in the lane, an ornithological paparazzi, tripods set, lenses poised, voices raised. It stayed all morning, in the spinney down the road, calling, posing, hiding. There are even photographs of the bird, while I took photographs of the throng. I was almost more interested in them in the end. I've never witnessed a twitch, and it was curious to have conjured one so suddenly and unexpectedly on my doorstep.

There have been other records of *Oriolus oriolus* in the shire, but none reported soon enough for anyone else to see. So the oriole made the day of many others, excited, chattering, grinning bird fans brightening that dreary June morning. 'It's a bird you dream of finding in your own patch,' one of them whispered to me, euphoric. I felt a swell of pride, for the old patch.

I once wrote about falling asleep in a deckchair in Italy as a Golden Oriole fluted somewhere above and beyond me in some poplars, lulling me to sleep, to dream. Last Sunday felt like waking from that dream to find the bird actually here. A fitting visitor for a Golden Jubilee, and failing that a Diamond. Would that more of our spinneys were so jewel-encrusted.

JULY

The sun is blazing down today, a rare respite from the wettest summer for a century. We seem to have entered an era of weather extremes. To celebrate I walk the six miles or so to work, arriving sun-kissed of cheekbone, yet soaked to the knee, and refreshed.

In the evening I receive a long overdue note from my Catalan friend Toni, an artist who illustrates wildlife guides, a man who has seen and even photographed a Goshawk chasing a Green Woodpecker through a pine forest. And all this from a hot-air balloon. He has more Gos action to report. "Goshawks have come into our lives ..." There is a photograph of a formidable young Gos, dark, as is the norm for the Spanish subspecies, with the resolute gaze that is universal. It is clutching a bait of dead pigeon, and Toni got this photo by setting up a trail camera with a trigger mechanism. The bird visited early. A year ago Toni lost one of his chickens to a visiting Goshawk, and now he keeps the chooks under cover. If he is up early enough he sometimes sees a Goshawk sitting on the church tower, looking for Collared Doves and Feral Pigeons. It reminds me that I should be up at dawn more often, when the average Gos probably fits a day's work into a few smash and grab raids.

Best of all though is the footage also enclosed with the message. Toni has been rearing three orphaned Kestrels and is introducing them to the world via the derelict tower. He has fitted a camera to the top of the building, to capture any action there, and as I press play, the image not of a Kestrel but of a Goshawk fills most of the frame. The bird has its back to the lens, though its face can be seen as it turns. It is mantled over something, which is of course hidden by its half open wings. The Catalan landscape extends to the horizon far below. The Gos is suddenly assailed from the left by a Magpie. The hawk shoots off, down, and immediately out of view. A couple of feathers drift out over the landscape. The Gos had caught one of the young Kestrels which, miraculously, survived the attack, as later in the day all three were present and correct in the tower. The Magpie had been an unlikely saviour.

Manuel Diego Pareja-Obrejon, falconer and author of the book about Spanish Goshawks *God Made You Eternal*, has a theory that the wide diversity he's encountered in Spain's Goshawks can be traced to the historic trade in the birds from all over Europe. There are Irish and Norwegian genes in these birds, and huge variation in young birds within a brood, he believes.

I have taken tomorrow off in hope of another warm day.

Friday

Inspired by events from Catalonia I wake at 5am to greet the dawn, squinting at daylight beneath the blind. I nudge it upwards with the back of my hand and rest the corner of the roll on my temple, to assess the day. No Goshawk, but the Sparrows are beginning to stir, in the Wisteria around the window. The Ash tree shivers in the gentle slipstream as planet Earth rolls towards the southern horizon and some wispy gaps in the dishwater grey cloud blanket. I hold on to the thought that at least it isn't actually raining, then notice the telltale ripples on puddles across the road. Then flecks against the windowpane. A French Partridge, bedraggled, picks at the grass verge close to the ribbon of puddles. Three Magpies steal into the scene – I have something in common with Toni's neighbourhood. They bounce to the puddles, to the roof of the barn, then away.

I had aimed to use today as a last chance to revisit T. H. White's old cottage, and to look for the Goshawk in the woods nearby. I've more or less given up on finding the Goshawk near here. But if I could find one there, that might make a fitting closure to this tale. I am running out of time, as far as Goshawk breeding season is concerned. Any young being raised will be on the point of fledging now. They will soon be gone, dispersed, unattached to place, quiet. Finding them will then become only a matter of chance, not of deduction. All this depended on warm weather. In wet weather the birds' spirits will be subdued. As will mine.

I'm up late, cataloguing Mum's old slides, now cleaned and scanned for her forthcoming 80th birthday. I have typed up the handwritten captions from each slide's frame. This includes a sequence taken in the Mediterranean when my parents took a boat home to Ireland – via Aden, Suez Canal, Etna, Portofino, Gibraltar – a year before T. H. White's ill-fated voyage terminating prematurely at Athens.

A Woodpigeon coos soothingly down the chimney, maybe using it to sound more alluring. "You *love* me, don't you?" as they might be crooning. If I can't hope to find the Goshawk today, perhaps I can find something else. I've been wondering if I should let Steve Wheeler know I'm visiting. It's been a while since he helped me find the cottage in the first place. I summon the courage to give him a call, to let him know, on the offchance he might have time for a quick catch-up. Three years have now elapsed since my first visit. At the time he had said he'd ask his aunt, who lives in the cottage now, and who knew Tim White all those years ago, if she would be willing to see me. I had taken the subsequent lack of contact as a no.

To my considerable relief Steve is full of enthusiasm for resuming this conversation. The papers and radio in White's day were full of talk of impending war. Today it's all about London Olympics preparation, and race rows involving England football captains. Steve has moved since the last time, a few miles uphill. When I find the new place he greets me like an old friend. He shows me the house he's been renovating and the land around it proudly, with its environmental features, its meadow, and the enormous pond that he's excavated. We discuss and trade books – I give him a copy of *Silent Spring Revisited*, and he lends me a book written by another of his aunts, about her side of the famous Edrich cricketing family. I tell him a bit more about my Goshawk investigation, and the place of T. H. White within this, and I share with him some ideas I've got about making a film about White's life, and commemorating the 50th anniversary of his death in 2014, perhaps by unveiling a simple memorial somewhere near here.

He does something unexpected. "I'm going to take you round to meet Jose," he announces. "She might not like it, but I really think we should try.

You can give her this book," he adds, tapping the copy of *SSR*. I can see he is apprehensive, and taking a bit of a risk, and I share his nervousness. But his mind is set, and we head over there in convoy. I follow him up the short drive and park in front of an outbuilding that is now on the site of the barn that White used as the Goshawk mews.

Steve knocks the back door while I loiter in the porch of a beautifully half-wild garden, full of flowering shrubs, climbers and border plants. Someone is obviously tending it regularly. He summons me through. I can hear him explaining to Jose that he's brought someone he'd love her to meet. She is regarding me with a mix of suspicion and shyness, and after some initial scepticism Jose slowly thaws on the idea, and motions for me to sit down: I have 15 minutes.

There is cigarette smoke in the room, which adds to the pervading sense of history. Rooms don't smell of smoke any more. She speaks slowly, clearly, in a refined voice. There is something impish in her fixed gaze, and contrary about her views, seeking reaction. Perhaps thinking of Tim recalls his manner. She lifts a pack of cigarettes from the table beside her. 'I suppose you are one of these people who thinks smoking is disgusting?' she asks, lighting up, testing my credentials. I shake my head. 'Would you like one?' she adds, hopeful. I nod. I can smoke a cigarette with my involuntary host, if that is what she would like. The air is so thick it would be peevish not to. It feels like part of the deal.

Jose has just had a birthday, and shows me a card someone sent her, containing a vivid crayon drawing of Tim, as his friends called him. He is smiling, bearded, sparkling, alive on the page. He now seems very present. I am able to grasp that I am talking to someone who talked to him, right here, 70 years ago.

White had rented the house from Jose's father. She is keen to correct the misgiving that the author was a lonely recluse living in a squalid cottage, as she had heard him described on a late-night radio programme. On the contrary, he had kitted the place out in deep carpets, wall-to-wall, gold-framed mirrors, candlesticks and thick curtains, even over the stone sink. "He was always putting on a dinner suit and going out," she recalls, smiling. "Or going to the pub on his bike, with a bar across the sit-up-and-beg handlebars for the bird, and one in the back seat of his Austin, for the same reason. He went to the pub a lot. He wasn't an expert on the things he wrote about, but he was a good teacher. He always made things fun and interesting."

She recalls, aged 13, helping Tim train the Goshawk. He would ask her to run with the twine, pulling bits of dead Rabbit behind her, afraid all the while that the Goshawk would ignore the bundle of fur in the grass and come instead after her. She feared its yellow eyes, she told me, much preferring the soft brown eyes of other birds of prey like the Kestrel. She's heard that Goshawks still survive in Scotland.

Jose relates a remarkable tale of being adopted one day here by a Kestrel, which she thought must have been an escaped bird. She resisted feeding it for three days, but relented when it continued to hang around. It visited for years, she said, and would sit on the fridge awaiting food. It brought fledged young to the garden, which would perch on the washing line, though she never saw its mate. Odd that Tim White's cottage should be singled out like this, by the bird. Did someone think he was still here, when they left it behind, I wonder.

She tells me about the birds around her, and the wildflower that grows just outside the window, though she cannot remember the name. I can identify it as a Willowherb of some kind – not Rosebay or Great, the two I know by name. I slide out from her shelves the *Concise British Flora*, and we find it in there, Latin name only. She seems pleased. She recalls Bee Orchids growing nearby, once, but not any more. I tell her this reminds me of Laurie Lee's account of drinking Bee Orchid wine with a witch. She invites me to look at Tim's paintings hanging in the stairwell, calling through to me to explain that visiting children found them scary. It's easy to see why, as they seem to represent his demons.

From time to time she reels in a remote control on a wire, and I wonder what she is going to turn on – perhaps the tv again – or off – me? It is her re-clining chair that is adjusted each time. I'm pleased to say that we have over-run on my allotted 15 minutes. Before leaving I sign her copy of *SSR*, careful to make sure it's Jose without an 'i'. "Have you seen this one?" she asks me, lifting another book from the table beside her chair. It's *America at Last*, the journal Tim kept on the US lecture tour that would be his swansong. "I didn't know about it till recently." I make a note to source a copy.

"I hope we will meet again," I say, as I'm leaving. I emerge blinking, from the half light of the cottage with its deep-set windows to the now bright sunshine of the mid afternoon. The garden is transformed and bedecked in colour. Steve has disappeared, left us to it.

I pause by the paddocks, trying in vain to befriend the horses there, in their cloth jackets. Jose thinks them pampered these days. My eye is drawn to the woods, to the north. A third of these, which Jose told me were known as the Crown lands, were lost to racetrack development from the old airfield that Tim White had known. I step carefully over ground churned by thou-sands of race-goers towed free from waterlogged temporary campsites.

With so many recent weekends rained off or otherwise unavailable, this might be my last chance to find the Goshawk. Imagine succeeding. Imagine Goshawk right here. That would be something to tell Jose and Steve. I think they'd like that. I am realistic about my chances. I have checked the latest available bird report for this county. It has this, for the Goshawk entry: "Once again among the more experienced birders in the county there is a difference of opinion on the status of this species. I have received indications of territory occupancy and/or breeding from seven sites ... While absence can never be proven, doubt has been cast on some of these sites as holding Goshawks. There were a number (six in all) of reports which do not obvi-ously refer to breeding birds. Of these, two descriptions were acceptable as only referable to Goshawks ... All suspected occurrences need to be fully documented by field notes to help establish the true status of this magnificent species in the county."

It doesn't offer much encouragement, but I search, through stands of trees where Tim also searched, three-quarters of a century earlier, for the bird that he had lost. I emerge from these trees and head across a wide open expanse of grass plots and grid-iron tracks, a row of flag poles pinging in the wind, rally cars roaring and throwing up dust on the track beyond a high fence. Tim liked fast cars and he learned to fly, although afraid of planes. But he feared for the Hobbies that lived in the woods here, and the imminent development of this race track. A lone dog walker puts up a thousand gulls dining on worms turned up by the churned mud and summer rains.

I enter the woodland, where another dog walker is calling forlornly for a lost dog called Willow. It's Forestry Commission land, wide, hard tracks, dense trees, wood piles with signs asking you not to climb on them for safety reasons, puddles. I feel gloomy in places, futile, and cheered in sunny interludes, especially when I find an unexpected White Admiral butterfly, the first I've seen in years, and orchids at a junction in the tracks. I do find a hawk. It rises quickly from the side of the track up ahead of me and flies across and into the trees and is gone. I think Sparrowhawk, and search the area from which it rose, in search of any prey remains. There isn't anything.

Further on Buzzards complain about me, and bounce off the flimsy tops of spruce trees to seek peace elsewhere. I find a remoter corner of the forest where the trees are Goshawk scale, and widely spaced. I pick a dry log to sit on and rest a while taking notes, to see if anything develops. I flick through a copy of *The Goshawk* I've brought. A Muntjac barks nearby, within the bracken. Buzzards plead for food from a nest some way off, and a raptor shape appears, through the trunks, lands on a pine branch. I have to lean forward to get it in view. Buzzard. It hasn't seen me. it flaps on after a few minutes. There is a motorway between me and the other part of the forest, and no sensible way of reaching it. I ache a little. It's time to make my way back to the cottage.

Back in the corridor between the stands of trees, I reach the cottage, not wanting to stand and stare, but now aware at which window Tim sat writing his book into the night, perhaps with a dram, and his beloved setter Brownie for company.

What became of his Goshawk, I wonder? Maybe it starved within days, or fell foul of a keeper. I like the theoretical possibility that it might have lived to a good age, wandering, looking for a partner in spring. Perhaps it was still at large when *The Goshawk* finally saw publication 15 years later. Even wild Goshawks have been known to live this long.

But if there is a Goshawk of any provenance here now, I can't find it, and with the summer now ending, it feels as though my chance of finding them, of ending this tale with a happy, even Disney-esque finale, has probably gone.

Chapter 11

ANIMUS REVERTENDI

AUGUST

A young Sparrowhawk has become a regular visitor to the garden. I've seen various of its inexpert, sometimes desperate and even reckless attempts to make its way in the world. It occurs to me that all across the land – or the parts of it where Goshawks survive, at any rate – young Goshawks are doing something similar. Consider the figures. If there are, as is currently estimated, up to 500 pairs of Goshawks nesting in the UK, averaging somewhere between two and three fledged young per pair, then somewhere between a thousand and fifteen hundred young Goshawks are jumping off their nest platforms, learning from their parents that food comes mostly from somewhere beyond the wood, and setting forth into the world beyond their natal homes. Let's go for the bottom end of this range: a thousand young birds. About this many have fledged annually since the 1990s, when the Goshawk population was first calculated to have reached 400 or more pairs.

On top of that 1,000 young birds newly produced each year, there are 800 to 1,000 parent birds, also roaming, some more than others, but by no means especially or necessarily attached to the nesting area. They can go where the food is; which, when food means crows of all kinds, pigeons, Rabbits, rats, thrush species, gulls, squirrels, ducks, Pheasants – means pretty much anywhere. These prey sources (or biomass) can be measured by the ton in our landscapes. Assuming a reasonable survival rate of hawks, in this food-filled, vacant environment, the Goshawk population might have expected to recover to thousands, if not tens of thousands of birds today.

I had hoped to have the science to back up the theory. I am not a qualified scientist – I got out of science after less than a year of it at university – but I have spoken to a number of colleagues and friends who are. I have sketched out with them how we might approach the scientific paper that could support (or otherwise) the theory. In the end, there hasn't been time to do it, because there are higher priorities than crunching Goshawk numbers to work out what probably happens to our Goshawks. But I hope that this may change, that there might yet be time found, that someone will take up the cause. I might get my name on the paper but in all honesty that isn't important to me. I would be content just to see the work done by reliable authorities.

Some colleagues have suggested the following approach: modelling suitable available places for Goshawks and prey to get an idea of what the limits might be on numbers. Correlates, I think they call them. But my sense is that this would be based on an old idea of Goshawks: the myth – for I believe it is a myth, an artefact of our own making – that they are birds of a particular kind of habitat; namely vast, remote, coniferous forests – the places in which they currently survive, which isn't the places that they need to be, so much as the places where they are *permitted* to be.

Because Goshawks aren't necessarily birds of remote forest. Like many species of higher-end predator they are simply birds that need food, and not to be killed; a tree to nest in, even sometimes in an isolated clump. If we can leave ourselves aside for a moment, Goshawks don't have any regular enemies here, when they get to adulthood; nothing that can easily catch and subdue a healthy mature Goshawk. In the absence of the Eagle Owl, they are the top avian predator of all but the mountains, where the Golden Eagle holds that title. Not much is going to kill them, except their own desperado behaviour. I have no difficulty believing that they are regularly, suicidally reckless, especially when young. But the evidence of their likely mishaps is a little thin on the ground. We would find more of them. The same with disease, to which they can be susceptible. We would find the bodies. I suspect they are dying somewhere else, unreported.

My instinct is to look at a neighbouring country – namely the Netherlands, just across the North Sea – with a comparable environment and situation; to look at how the Goshawk population is faring there, relative to numbers of other species such as we also have here – raptors, primarily – then work out how many Goshawks we might reasonably be expected to have.

The Netherlands is largely suburban. Like our own situation, its rural areas comprise farmland with scattered and fragmented broadleaved and mixed woods. To all intents and purposes it resembles lowland southern England. Perhaps my own part of England is the closest to it in terms of geography and appearance. In Britain, both Sparrowhawk and Buzzard are around *a hundred* times as numerous as the Goshawk. If we had similar ratios to those in Netherlands raptor guilds we might expect to have thousands of Goshawk territories, rather than several hundred. There is some secrecy about Goshawks nests in certain areas, but that doesn't explain the vast gaps in distribution.

Clearly these are 'back of a fag packet' calculations., but I have done another simple bit of arithmetic. Goshawks produce almost as many young as Sparrowhawks. Although they are very large birds, they are sexually mature and can breed at just one year old. Assuming a survival rate among young Goshawks comparable to that of young Sparrowhawks, and that there really isn't a habitat constraint on the species, how much might we have expected the population to grow since it reached 400 pairs in 1994? Growing at the rate of just one new Goshawk per pair, per year, it increases like this: 400 pairs, 600 pairs, 900, 1,350 ... etc.

Even if these figures are optimistic, the question is worth asking: why hasn't the Goshawk population grown *at all*? Another question, almost as intriguing: why is no one bar a handful of dedicated, knowledgeable and courageous raptor conservationists asking the question?

I can't be sure of the answer, but what is clear is that hundreds and quite likely thousands of young and some seasoned Goshawks are disappearing every year, without trace. Where can they be going? The bodies of large birds, and especially birds of prey, are eye-catching. You don't have to be an expert to notice them, or to think of reporting them or even bagging them up and handing them in (although it is always safer not to touch, and to take photographs and submit details of location instead). A large proportion of these young Goshawks have rings on their legs, mostly put there by Raptor Study Group volunteers: all the more reason for someone to report the birds and/or their rings when found. The rings have a code and address on them.

"For all the time I have been ringing Goshawks, only two rings have been recovered," Malcolm has told me, from his Scottish Borders study area. "I have ringed 486 Goshawk chicks. One bird [that was recovered] had been shot and I suspect that had it been shot dead then picked up by the shooter it would have been disposed of, and no ring recovery recorded. However, what is likely in this case is that it had been shot and wounded, then managed to fly away, before dying. The other ring recovery is still under investigation."

After all his tireless years of record-keeping and devoted searching, Mick has reviewed population statistics for the Goshawk in north-east Scotland. He's been looking out for the Goshawks there ever since he found the first birds nesting in the early 1970s. He knows the life history of the population in the entire valley, nest by nest. His analysis shows that the birds survive and are able to nest in state-owned woodland. This fact is well established. What he can also show beyond doubt is that Goshawks do not survive well when they disperse beyond these sites. In fact, were it not for the birds hosted on publicly owned land, the overall population would not be sustainable. In short, the birds would die out again even in that region.

This is just one study in one place, but it also happens to be one of the major Goshawk strongholds in the UK. It seems reasonable to deduce from Mick's work that without state-owned forest UK-wide, we could well lose the Goshawk again as a breeding bird in the UK. Another extinction, 130 years after the Victorians, untrammelled by wildlife protection law, first killed them all. It doesn't bear thinking about, in an era of government obligations to biodiversity conservation, and setting an example to the world on saving species. It may also tell us about one of the potential implications of any potential sell-off of the nation's forestry estate to private interests.

Friday

I'm used to Pheasants playing chicken on local roads, especially in autumn when 35 million of these hand-reared birds are released from nursery pens countrywide, to take their chances in the 'wild'. I have just had a narrow miss, this time with a male Pheasant. He was standing in the road, catatonic. I had time to stop, get out, take a box from the boot and lift him into it. He looked alert, unscathed; in pristine condition, even. Like most of his kind this is a hefty bird, weighing about a kilo, with stout legs for sprinting and raking. But he made no complaint or struggle. I guess he has been dealt a glancing blow by a passing vehicle and was concussed. He's been lucky. Three million or more just like him are not so lucky in their vehicular collisions each year. That's a lot of collateral damage, and not just to the birds.

It occurs to me that I've not really been this close to a live Pheasant, or properly studied one. These are spectacular birds, with plumage, as John Ruskin described it, "inlaid as in a Byzantine pavement, deepening into imperial purple and azure, and lighting into lustre of innumerable eyes".

Dusk is gathering, so Fester (he needed a name) is kept in overnight for observation. Perhaps he'll perk up by the morning.

Writing a hundred years ago, W. H. Hudson had a name for the Pheasant. He called it the 'sacred bird'. He did this, he explained, 'to express condemnation of the persons who devote themselves with excessive zeal to pheasant-preserving for the sake of sport".

He detested a Pheasant in what he called "the preserves". "One odd result of this over-protection of an exotic species and consequent degradation of the woodlands is that the bird itself becomes a thing disliked by the lover of nature." But his view changed when this Asiatic species earned its place in our great outdoors, when "the sight of him affords me keen pleasure".

I have read that game birds like these aren't categorised as livestock, which means that they aren't subject to the same legal guidelines as are in place to protect chickens. Nor, technically, are they property, I think because they are routinely released into the environment and can wander – these birds hand-fed in confined spaces since hatching – where they like: into neighbouring estates, dual carriageways, this garden. Nor are they wildlife. It is illegal to release non-native species into the wild.

Saturday

Early this morning Fester has had a sudden moment of lucidity, recognised his situation, seized his moment, and bolted for freedom from the basket while Sara is taking his picture. We now have a photograph of him hot-footing it across the garden pond in a bee-line for the hedgerow. We hope that he is still out there, integrating in a way that might have pleased Mr Hudson.

Sunday

We get a visit from a Peacock which lives on a farm down the road. I watch from bed as he struts down the field margin opposite, swishes into the paddock through the open gate, climbs on to a fence post, and gives his foghorn 'cah-haarrk' (get me!) in the direction of the spinney, no doubt hopeful there might be Peahens in there, of course oblivious to just how far from home he actually is. This has all the hallmarks of a land grab. Of course no Peahens respond, but a male Pheasant not unlike Fester appears, clearly startled, from the hedgerow beyond, and promptly legs it down the street, Roadrunner style. What must the sudden arrival of a shrieking Peacock look like to a Pheasant up to now thinking itself the cock of the block?

Not much has been done to assess what impact on the ecology of these islands the annual release of 35 million Pheasants might have been having. Add to that another five million French Red-legged Partridges. Pheasants alone make up a third of the land bird biomass of the UK. There is no licensing system for shooting estates so no need to apply or qualify, or to worry about losing the right to put Pheasants out. Nor is there any obligation to protect the birds from airborne predators by putting roofs on enclosures. A tall, stout mesh fence dug a few feet down might be just about enough to protect the Pheasant poults from Foxes and other ground-living threats, but of course it won't keep wild birds out. We know that a Goshawk can make short work even of adult Pheasants, so it's not hard to imagine what they must make of a herd of flightless immature Pheasants cornered in a thicket. Is it humane to ignore the consequences of setting up this scenario in woods and spinneys the length and breadth of the land?

One gamekeeper has told me that nowadays there's little work for him to do on killing Magpies and Carrion Crows. His dealer rears the Pheasants to sufficient size that by the time they reach the open air they are beyond the scope of corvids. They do cost £3.50 each, though, and the husbandry involves simply putting these well-grown poults into the pen and feeding them for a few weeks, as might be done for any other poultry. The Pheasant-shooting season begins at the start of October. Another keeper reports that the presence of Buzzards can actually be helpful, as they aren't particularly adept at catching well-developed Pheasants, even cornered, institutionalised ones. Their soaring presence is thought to be enough to keep the Pheasants – when released – herded under cover of the wood, and to discourage them from straying into open country. He didn't mention any direct personal experience of the Goshawk. I didn't press.

I'm not the only one who's been taken by surprise as the government announces it is going to fund a study that would involve destroying Buzzard nests and taking some of them into captivity, to see whether this might result in even more Pheasants being reared in these open-to-the-air pens. Hard to believe there is a need or justification for legally culling birds of prey like Buzzards in order to preserve a few more Pheasants to be shot at, if not run over. And if this is their attitude to Buzzards, what does this say about the prospects for the returning Goshawk? I wonder at the relative cost of this research, versus the cost of putting mesh roofs onto release pens. If the Government were testing the water with this announcement, they were rapidly blown out of it by the vehemence of the public backlash.

On the point that it is illegal to release non-native wildlife into the environment, I'm not sure how the Pheasant industry circumvents this basic law, every year, on such a scale. If we are to continue to tolerate the manufacture and shooting of Pheasants at its current, unlimited level, perhaps we should be calling for better regulation, if only for the well-being of the Pheasants themselves. While we may have become somehow desensitised to them through their apparent domesticity and abundance, like any animals they deserve respect, and humane conditions. The mistreatment of animals is ultimately degrading for us all.

If it could become a basic requirement of Pheasant and Partridge rearers to ensure young birds confined in pens are protected from aerial attack, there would be less pressure on keepers to kill or mistreat avian predators and the decoy birds and other live baits they use to lure them into traps. I take no pleasure in seeing a modestly paid working individual criminalised or even jailed for trying to do the job expected of him by their boss. If all release pens were required to be secured, and numbers of poults limited, they would be easier to inspect, licence, and close down if standards aren't met. Landowners would have to make the funds available for roofing. This would create rural employment. Young Goshawks and other raptors and owls would have to forage elsewhere, and learn how to catch other prey, like Woodpigeons, squirrels, corvids. Estate workers would keep their jobs.

* * *

"The great soaring bird is nowhere in our lonely skies, and missing it we remember the reason of its absence and realise what the modern craze for the artificially reared pheasant has cost us." It's W. H. Hudson again, all those years ago. He'd be pleased, I'm sure, to know that the Buzzard is more or less back where it belongs. But he might also wonder why our Goshawk isn't yet right here with it.

"The soaring figure reveals to sight and mind the immensity and glory of the visible world," he adds. "Without it the blue sky can never seem sublime."

On the legal ambiguities surrounding Pheasants I have checked with my barrister friend, who specialises in property law. It has crossed my mind I might have been poaching, when I brought Fester home. My friend has consulted the relevant literature.

"I think I can address the point on ownership," he writes back, in due course. "The case law concerning ownership of animals in this area is old and the language archaic. The ownership is described as 'qualified' as the ownership can be lost if, for example, the animal escapes and has no *animus revertendi* (intention to return)."

I don't know if Fester had an *animus revertendi*, but I'm pretty sure the Goshawk does. It belongs, in the end, to all of us, and to none who has the right to kill it. It can go where it pleases. I envy it that.

SEPTEMBER

Work brings me to Germany. The pharmaceutical firm Boehringer Ingelheim has created a drug called meloxicam. It is the only non-specific anti-inflammatory drug we know of that doesn't kill vultures when they eat it with the tissue of dead livestock. The Boehringer board has also given up the patent on veterinary meloxicam so that other companies can produce it, and they've provided some funding for the wider effort to save the birds from extinction in Asia. I'm here in Germany to give a presentation on the project's progress so far, and set out what needs to happen next to put vultures back into the environment where they belong. We have vultures in a captive breeding programme and bans in place on veterinary diclofenac in India and neighbouring countries. So far, so good. But saving species isn't cheap. The meeting goes well. I even manage to squeeze a photo of the Goshawk in there, and some discussion of the Habicht, as they call it.

Walking to the train station beside the vast, high-tech campus of the company's headquarters I board a crowded train late in the afternoon for a three-hour journey north that hugs the west bank of the Rhine. We follow the languid slide of the wide river through wooded hills scored with vineyards. The occasional Buzzard drifts south on migration, against the flow. My destination is Cologne, one of the handful of German cities that the Goshawk has reached, and in which it evidently thrives. This is all I know. If I am to find the Goshawk here, I am on my own; it will be without local knowledge. I like the idea of applying what I've learned, trusting my instincts, although not sure where I'll start: the green bits on the map where perhaps it rests (and nests, although not in September), or the streets where it hunts?

Goshawks aren't the only reason to visit this city of a million inhabitants, of course. Cologne's cathedral is a site of pilgrimage. In fact it's the most visited landmark in the whole country. It was begun in the 13th century and completed in the 19th, surviving dozens of Allied bombs during the war, as the city around it collapsed, and patched up soon after. The visitor by train is confronted by this gothic mountain of a building at the moment of emergence from the terminal. The Jackdaws around it are as moths to a building of conventional scale. It is jaw-dropping, ornately carved yet organic in feel, like a fluted, sooty termite mound of the gods. Gazing up at it, among the ranks of the open-mouthed, I wonder if the Goshawk ever visits it too, to hide amongst its sinister cloisters. I can check it again tomorrow. Right now I have an antik hotel to find in what remains of the old city, and the daylight. I can look for non-Cathedral Goshawks on the way. I am struggling to find any birds at all, I realise, after my eye is drawn to two pigeon shapes on top of a building – only to establish that they are fakes – then an outsize dove with a bright green olive branch over the entrance to a church. It dawns on me that there are no real pigeons to be seen.

After checking in to my room on Kaiser Wilhelm's Ring I explore the park that it overlooks, and walk for a couple of kilometres in the general direction of the old town and river. In the two hours of daylight remaining to me I don't find a single pigeon, in places where they'd be so familiar in other cities you probably wouldn't even register them. I investigate the squawks of a Blackbird, over the clanking of a tram. But I can see neither it nor the source of its disquiet. The Cathedral is even more breath-taking in the white, angled night-lights against the black sky. Bats flit like fireflies around its monochrome cliffs. The river bounces the sound of trains crossing the iron bridge, rumbling like thunder. Nocturnal gulls wheel around the great arcs of riveted iron. The pigeons, if there are any, must be fast asleep.

Saturday

I pull open the shutters and scan once again the tree-lined square outside through light rain. It's bigger than a football pitch, with 25 mid-aged Plane trees laced through it, a water feature and neat lawns, benches, shrubs and flower beds. Still no pigeons, though I can hear against the wet traffic the complaints of tits and finches ringing from within the foliage of the Planes. I set off before breakfast to explore the parks that ring the inner city. Still no pigeons on the street, but within five minutes I spot familiar puffs of them issuing from the roof tops, and Woodpigeons commuting across the skyline. Of course there are pigeons here. I enter the first of the parks, rain dripping loudly onto the first shed leaves of autumn. It remains conspicuously birdless at ground level, although there is activity among the branches. A pair of Blackbirds forages in tandem, a safe distance out from the park's leafy periphery. I cross a road and discover another, larger park and, in the middle, I find the birds: a flock of grazing Woodpigeons, sprinkled with crows, Magpies. I watch a group of ten Blackbirds, one Fieldfare with them. I can hear Mistle Thrushes and the odd Song Thrush, and a Red Squirrel picks its way through some parked cars. A lady leading a small dog in a jacket tips crumbs out of a polythene bag, and a throng of birds follows her. The curious thing is that they aren't Feral Pigeons, they are Carrion Crows. I count 40. The Feral Pigeons are still sticking to the high tops.

After mid-morning breakfast I walk another couple of kilometres to the cycle hire shop, in the bowels of the road bridges. There are no bikes left. I'm already leg-weary but must continue the search on foot. I follow the Rhine north again, through flea market stalls on one side and driftwood on the other. I wonder if I can find a Goshawk among these antique sellers, and I strike lucky: a carved rack of coat hooks, in the form of a hawk. I can tell it's Gos because of the orange beady eye glued onto it. It would definitely spook anyone removing a coat. I walk till dusk, but this is the only Goshawk I find.

Saturday

Back home. I've uprooted some ripened Teasels from awkward spots in the garden and slotted them into the hedgerow, so they protrude like a candelabra. Within hours Goldfinches have taken the hint, illuminated in the dawn sunlight, working the bristly cones to extract seeds. Frumpy sparrows perch nearby, as usual, admiring the glitzy newcomers and their craft. Migrant warblers are passing through, feeding up as they go. A Chiffchaff works the Weeping Birch energetically, pirouetting in pursuit of flushed gnats.

I've collected two loads of off-cut wood from a neighbour on the last two evenings, which I have to sort and cut for firewood, or practical uses in some cases. She keeps hens, and has noted how they flinch when a Sparrowhawk is overhead. Interesting that these fowls even notice, and care. These are big chickens. There's just something in the appearance of a hawk, I suppose.

There's that famous experiment which found that newly-hatched chicks cower when a hawk-shaped silhouette is drawn across the sky above them. Draw the same silhouette the other way and it looks like a goose. The chicks don't react. My German pharmacist friend Juergen has told me he would die happy if he could find a biochemical explanation for this kind of in-born wisdom.

There's been a noteworthy development that will be of interest to anyone who ever misidentified a bird of prey. Some people think they have observed Hobbies and Sparrowhawks mimicking the flight patterns of other species, in order to approach closely enough to make an attack. Patrick Stirling-Aird has written to the journal of the Scottish Ornithologists' Club and cites Jack Mavrogordato, one-time leading falconer here, who must have been able to read raptor flight better than most, who had perceived this behaviour in a Sparrowhawk (flying like a Lapwing) back in the 1970s.

I am thinking back to the occasion in late summer where I thought that a Hobby, approaching high above a flock of jittery hirundines, was itself a Swift, some weeks after the last of the Swifts had departed, hence me giving it a second look. I've also seen a Sparrowhawk approaching a garden in the casual, breezy manner of a dove, before going suddenly into attack mode. I consider all those other misidentifications, or uncertainties. Perhaps I can now excuse myself on the grounds that raptors may actually be impersonating each other, or even non-raptors. 'Perhaps I'm off the hook then, Mick?' I ask the expert, hopefully.

Sunday

I'm sitting in the back garden deckchair reading T. H. White's journal of his US lecture tour. It's another rare high-pressure day after a summer that never really got going. The Swallows have thinned to a scattering. A darter dragonfly joins me, perching on the top of the page a centimetre from my thumb, feet hooked over the edge, head tilting, red chilli-coloured abdomen pulsing, four clear-veined wings angled forward. It launches forays over the pond, and returns, determined to share the book, hovering close even while I'm turning a page. Above me the dying, bare-branched Eucalyptus tree is shedding its supple bark like a banana skin, strips dangling to reveal the smooth, pale sapwood beneath. It has become the local watch-point of choice for birds of all kinds. I counted ten species in it the other day, including a Yellow Wagtail being oddly delinquent, getting psyched up for the flight to Africa.

In a moment the tranquillity is broken. A Blackbird screams dramatically, even by its usual neurotic standards, from the outermost twigs, and Collared Doves clatter in a panic out of the crown, shedding breast feathers on the way. I am struck by the extremeness of the response, particularly given these birds can't exactly have been taken by surprise, given their vantage point. My view is obscured by the fruit trees and pines so I place the book, pen and notebook on the grass and ease myself out of the deckchair to slip quietly through the young trees to the fence at the bottom, for a look round. I can see nothing, out of the ordinary, trace no source of disquiet, although a nervous murmuring persists in the shrubberies of neighbouring gardens. The gulls that have been a regular feature of the ploughed and already planted wheat field at the back are conspicuously absent.

I return to the book, the warmth of the afternoon sun. A short while later I am distracted again. All too readily. There is a row going on among the Rooks and Jackdaws at the end of the ploughed field at the end. They have risen from their toil and are swirling around over the oaks, close to where the Buzzards nested over the summer. Adjusting my focus from two feet to three thousand, I sift through them, checking if there is a raptor profile in the midst of the ragged melee. Again, no marauder; nothing untoward.

I re-open T. H. White, and his evocative snapshots of the USA of 50 years ago, the era of Kennedy rumours, civil rights movement, *Silent Spring*. At the time he wrote *The Goshawk* he never struck me as a bird lover, as such, but by 1963, 25 or so years later, he is clearly entranced by them, among his many interests. I put pencil marks against the regular bird and other nature observations he records as he makes his way across the States. I am wondering if I'll find a Goshawk among them, not that he would have been likely to encounter one, even in what turned out to be a strange year of major Goshawk invasion from the north of the continent. By West Virginia he is regretting the lack of time to get into the wooded hills, where Shawnees once ruled, the only wildlife he's seen there being a road-kill possum in the headlights.

While I'm absorbed in this, a message arrives from David, my eyes and ears at the north road and river end of the village. He doesn't do build-ups, David. "A pair of Goshawks floated over us yesterday. I was holding bits of wood while Hobnob Dave screwed them into the fence. They were very close – 40 or 50 feet up – and stayed for a minute or so before gliding off towards the A1/Ouse."

Really? I'll admit it crosses my mind that David might be losing the plot, not least because he didn't call me about this yesterday. I call him back straight away. I figure there's no substitute for a first-hand account, and the sooner I hear it the more accurate it will be. It's all, in the end, in the detail. "I need to come round and take a statement," I joke. Actually I'm only half-joking, I realise, as I'm putting a little black notebook and pen in my breast pocket. I have a good look at Esme Wood as I pass, and minutes later David and his wife Juliet are walking me through the incident. She saw them too. Two witnesses. That helps. A promising start. David's scribbled draft report for the county bird recorder is lying on the garden table as we pass, to begin the 'walk-through' of yesterday's events.

They saw the first bird approaching from the north-east (it would have been visible from my garden, I note), not very high, Buzzard-sized but clearly not a Buzzard. They describe its tail being long and tilting, rudder-like. Swallows were on its case, sounding the alarm. All the other birds scarpered. It came right overhead – "just twice the height of that willow" – says David, and in a moment another one was with it. The raptors seem to have been of a similar size. The later bird was calling. Loud. '*Kyek kyek, kyek, kyek* ...' this fits hawk. The Buzzard family *mewls*. They were gliding over and around each other in an elegant ballet.

Walk-through complete, we continue to chat over a glass of wine in the garden. We've gradually stopped all talking at once. David has worked through all the other things it could have been, and ruled them out. I suggest another couple of possible passage migrants – Honey Buzzard, the harriers (a Pallid Harrier passed through close to here a year ago) – and we rule them out too. They have either forgotten or more likely I haven't told them before that I've been writing a book about the Goshawk.

I pass on some of what I know about how Goshawks disperse in the autumn, and why our local woods and fields could be as good as any, if a Goshawk were to get this far.

While we are discussing this, a Sparrowhawk appears overhead, twice, circling. The second time it folds into a diagonal stoop, aiming I would estimate at my end of the village, half a mile away. David and Juliet know their Sparrowhawks. Whatever they saw yesterday, it wasn't Sparrowhawks. David points out the Woodpigeon still relaxing on the neighbour's roof. It hasn't scarpered, this time.

David didn't put the word out at the time because he was too busy, and because he figured the birds weren't hanging around. I don't know if it has translated to the page of his report, but for me their story is gaining in credibility. The case for local Goshawk may be re-opening. Unless both birds were just passing through ...

Back home I go out on the flat roof, to scan the wide horizons in the evening sun. "It would make a great raptor watch-point, this," I call down to Sara. She laughs. The Eucalyptus tree may already be serving that purpose for pigeons, although it is currently unmanned. Scanning around, I can see that the ploughmen have been busy. Even Crane Hill and Hungry Hill to the north-east have had their several years of wildflowers and low scrub undone. To the south I can see gulls, 300 or more chasing the plough, bingeing on morsels, this side of the Enchanted Forest, wherein might lie some other, all-seeing eyes, following the gulls' progress.

I think we really might have had Goshawk here yesterday. David knows what it wasn't. I know I'd more or less given up hoping, or looking, locally, but maybe they are here now. One passage bird would be unusual, but two – together – here, of all places? I will have to put the book-writing aside tomorrow and have an overdue rake in the wids.

Sunday

David copies to me his carefully scripted official report to the county bird recorder, now typed up and including Juliet's contribution. This is an extract:

"Bird was not as big as a kite or buzzard but of sturdy build and not a harrier. It was joined by another bird of prey (appeared from nowhere) and it was initially thought that a scrap was about to take place. Both birds flew right over our heads (around 20–30 feet above us) and started a rather sensual display of gliding first above and then below each other which made the second bird more likely to be a mate rather than an opponent. As the two approached each other the second bird called '*kiaow, kiaow, kiaow*' (perhaps six or seven of them very quickly). The second bird was more browny-grey on the back, seemed slightly bigger than the first bird and had a more browny-grey speckling/scalloping on its more creamy-grey underside. It too had clear dark bands on its long tail, which it occasionally fanned out. Both birds performed this gliding 'dance' for 20–30 seconds before flying off to the north west."

I show this to Mick. "This description looks good," he replies. "It is spot on for timing – birds up in the air in a few places this month. It doesn't last long but is good to see. Not sure what it's about – possibly keeping dispersing juveniles from settling in places occupied by adults? Not sure; but it happens annually." For what it's worth I add Mick's endorsement to David's 'bid'.

Could a passing Goshawk have been being intercepted by an established one, I wonder? I walk over to the Enchanted Forest in the morning, picking my way carefully over the woodland floor, keeping near the edges, away from the paths, where a Goshawk might find a quiet corner. Before long I am finding pigeon feathers strewn around over quite a wide area. At first I think these could have come from a bird plucked in one of the oaks above me – that would count as Gos evidence – but the area becomes so extensive that I realise these are simply the moultings of a roosting multitude of Woodpigeons. I do find one dead, but it is barely predated at all, for some reason. Perhaps it was winged by gunshot and died here later, from its wounds.

Elsewhere I find pigeon remains in a couple of places, chewed by Fox. A Muntjac sprints away as a dried branch cracks under my boot. A Buzzard mews and departs. I find a particularly impressive primary feather from a Buzzard, and a juvenile Goldfinch dead on the path. I soon find myself pulling plastic mesh tree guards from tree trunks. I wish I'd got here earlier. Some of the saplings have been garrotted and stand rotted and dead. I could push them over, but I leave them upright for the woodpeckers to fell.

I stop by at Stonebridge Farm, to speak to John, the farmer there. I think this is the house in which the poet John Clare slept, in the porch, half-way through his walk after breaking out of the asylum in Epping Forest. Clare was destined for his home village, another 40 miles to the north of here, thinking in his confusion he might find his childhood love there, not knowing or believing that she was by now dead.

I arrange with John, the current farmer, a survey of part of the farm for later in the week, as 16 kittens swarm around our feet. He gives me some of the paperwork from the volunteer/farmer alliance surveys David and I have been carrying out on the land he farms. I tell him about the pair of White Storks that one of my colleagues spotted in his fields the other day; I'm guessing maybe Swedish birds stopping off for a snack on their long journey south.

I end up back in a corner of the forest where the front wheel of my bike came out of its forks a fortnight ago. I have a quick scan of the woodland path for the missing wheel-nut, and, to my astonishment, actually find it. I take some encouragement from this small victory. If I can find a stray nut in a forest, then there remains hope on the Goshawk front too.

Chapter 12

AMERICA AT LAST

Among the scattered Victorian records of Goshawk in Britain I've come across an unusual one. This was a bird shot in May 1869 on the flank of a mountain called Schiehallion, in Perthshire, not far from where I once lived. It is a place I know, a hill I have been on. The name may be an Anglicisation of the Gaelic 'fairy hill of the Caledonians'. Or, perhaps less poetically, it might mean 'maiden's pap', or even 'constant storm' – the latter perhaps a candidate alternative name for the Goshawk itself. But what intrigues me most about the record is that this Goshawk was apparently one of the North American subspecies. How on Earth did that get there? Or is it, in fact, just another a fairy story?

The Goshawk in North America is generally regarded as not one but two (some say three) separate subspecies: in the east they are *Accipiter gentilis atricapillu*s and in the west *Accipiter gentilis laingi*. The birds that inhabit the Rocky mountains as far south as Arizona are also known as *apache*. These American subspecies have a different appearance to those of north Europe, *gentilis gentilis*. And this odd record for the Perthshire mountainside in the late 19th century isn't the only one for the British Isles. England has one record, a bird that arrived in 1931 in the Scilly Isles, that magnet for lost birds and those compelled to go and see them. There is that handful of reports for Ireland too, where Goshawk records of any kind have been scarce since its extinction as a breeding species two centuries ago.

It is remarkable that the Goshawk might be able to cross the 3,300 miles of Atlantic Ocean. I think we know that their European relatives are not great crossers of the relatively short stretch of the North Sea, although they can and will do it. Goshawks are powerful birds, clearly, but they are built for sprints, and carry a lot of bulk and muscle, not to mention the hardware in their formidable feet. They do soar, of course, but not routinely. I sense they do not rest all that easily on a thermal. A female Goshawk weighs three times as much as a female Hen Harrier. She is half as heavy again as her Buzzard counterpart. She is not called a short-wing for no reason. The Goshawk is not built for long periods idling under the stratosphere. An Atlantic crossing is more than a marathon for any bird. I wonder if ships must be involved in each case in these apparent trans-Atlantic navigations. I also wonder if the National Museum of Scotland still has the Perthshire specimen.

It feels like a long shot, but I get in touch with Dr. Andrew Kitchener, Principal Curator of Vertebrates. "I wondered if you could confirm the record and if you still have the skin at the Museum?" I ask him.

"We do indeed have the specimen, which is a mounted skin," he replies. "My colleague Bob McGowan has done a lot of research on the origins of this specimen, and will be able to tell you a lot more about it."

Bob is Senior Curator of Birds, and duly obliges. "As Andrew has indicated, the Perthshire specimen is still extant. I spent some time investigating this occurrence a few years ago when I was on British Ornithologists' Union Rarities Committee (BOURC) and it was reviewing a number of Category B records. You ask if I can 'confirm the record' and I can certainly affirm that it is an example of *atricapillus*.

"The record has a somewhat chequered history," Bob goes on, "originally having been accepted on the British List and accepted by the main 'establishment' ornithological authorities of the day. Latterly it was unceremoniously dropped from the List (in the BOU's 1971 *Status of birds in Britain & Ireland*), though apparently without any formal review.

"There is a fairly substantial quantity of background information on this occurrence which was never published. Anyway, the BOURC has the file now and I believe it is still circulating. My understanding is that the review (whatever the decision) will be submitted for publication in due course (normally to *British Birds*)."

In light of Bob's advice, I call Dr. Martin Collinson, my colleague on the *British Birds* board and Chairman of BOURC, for a chat. Martin confirms that the file is currently being assessed. It seems the American bird spent a couple of weeks in the premises of a Glasgow taxidermist, where some confusion or mix-up cannot be ruled out. However, it is generally accepted by the experts that the Goshawk *can* make the Atlantic crossing (half a dozen Irish records and a Scilly bird are hard to ignore), although the help of passing ships cannot be ruled out – not that this would invalidate the navigational feat by the bird.

Perhaps a hoax cannot be ruled out, but who would go to such trouble to place an intact American Goshawk on a Scottish highland mountainside, or give it to a Glasgow taxidermist without realising its genuine, exotic import value?

I have looked up the original record of the find in the journal *Ibis*, from 1870. Robert Gray recorded it thus in his letter to the publication:

Glasgow, 4th February, 1870

SIR, – Having been engaged for some years in the preparation of a work on the Birds of Scotland, I have personally made particular inquiries throughout almost every county regarding the occurrence of the rarer species, and have been rewarded by the acquisition of many facts of interest, which I have no doubt will serve a useful purpose when they are published.

Among birds of this class that have lately come into my hands, I find about half a dozen species that are not mentioned in the last edition of Yarrell's 'British Birds'; and as two of these possess an additional interest from the fact that, being nearly allied to birds already known as British, they may have been overlooked and are likely to occur again, I beg your permission to put them on record in 'The Ibis.'

Last May, when at Brechin, in Forfarshire, I was fortunate in procuring a specimen of the American Gos-Hawk (*Astur atricapillus*) which had been killed in the vicinity of Shecallion, in Perthshire. It was sent by him, along with a number of Snow-Buntings and other birds, all recently skinned, to the person from whom I got it; the specimen having been very roughly prepared, as, on afterwards preparing to relax it, the Glasgow bird-stuffer, whom I employed to mount the skin, found that the brains and eyes had not been removed. This specimen, which is an adult, and apparently a female, is 24.5in. in length; the wing from flexure measuring 14 inches, and the tail 10.5in ...

It is possible, therefore, that, like other North-American birds (as for instance the Gos-Hawk just mentioned) bearing a likeness to British species, it may come oftener to this country than collectors are in the mean time aware of.

I am, Sir, your obedient servant,
ROBERT GRAY.

Dr. Collinson tells me that the Irish records are also under another review, by experts there. Their findings may have a bearing on the acceptability of the Scottish record. If accepted, they would at least ease some of the doubt over the birds' ability to make the journey.

I've been thinking a lot about the pirate hawk's journey. It is 3,223 miles from New York to Glasgow. Consider those mid-Victorian days when masted sailing ships needed two or three months to complete the voyage. Even steamships would take at least two weeks. Any stowaway Goshawk would have to find a way of nourishing itself from prey on board – they can't go many days without eating – and any lost Gos would have worked up quite an appetite if lost at sea, battling to stay clear of the waves. Perhaps, once on board, it might tuck itself away in the rigging, and catch ship rats, although rats of any kind are largely nocturnal. Perhaps it could ambush seabirds following the boat. In fact when I check on this with Rob Lambert, environmental historian, he cites the example of a Red-tailed Hawk on board a boat from the US that was observed doing just this. However the US Goshawks reach our shores, at face value the achievement is astonishing.

I also get in touch with Kieran Fahy of the Irish Rare Bird Club, to find out if the records from there are likely to be verified. "There are six Irish American Goshawk records," he says, "five of which fall within the remit of the Irish Rare Bird Club. I can confirm that it is currently planned to review the Republic of Ireland records but this will not be done until some of our current reviews are finished. I am in the process of accumulating such details as I can on the records but for some of them, this is proving difficult."

Kieran also lets me know that the single record for Northern Ireland, from County Tyrone, falls under the jurisdiction of the Northern Ireland committee. The specimen still exists.

T.H. White's second and last visit to America coincided with the Goshawk invasion of autumn/winter 1963. Not that he, like most people, may have known much about it, but the coincidence is pleasing. His lecture tour took him through New England and the mid-west to the west coast, and he kept a journal of his impressions, apparently without intention to publish. Like me, he "thought New York was stunning – not terrible, as expected".

It was on Jose's recommendationt that I duly ordered *America At Last: The American Journal of T. H. White*. The book was never published in the UK, and is now out of print. My former library copy is despatched from the US. It provides a vivid series of snapshots of the great continent at a time that Tim considered a major cultural renaissance. Although at times exhausted by a punishing schedule, he declared that he had never been happier.

His love of and interest in wildlife is undiminished, and observations of the many exotic species he finds in the course of his tour punctuate the text throughout. And although he doesn't find a Goshawk he does note encounters with Red-tailed and Swainson's Hawks. His health is generally robust on the gruelling trip, and he is teetotal throughout. That said, he is in hospital for treatment on severe abdominal pains when word reaches him that President Kennedy has been assassinated. The shot that rang round the world prompts Tim to reflect on *The Goshawk*. Intriguingly, someone once told him that George VI seems to have been reading this of all books the evening before he died, as it was found on the bed when the King was discovered dead there on a February morning in 1952. I haven't been able to verify this, and I wonder if it is recorded somewhere within the Royal family's archives, or their collective memory.

Tim had been in the US just once before. When the stage version of *The Once and Future King* was shrunk down into the musical *Camelot*, he travelled there to meet the cast and crew, headed by Julie Andrews and Richard Burton. The charismatic author proved a big hit with all, and was forgiving of the surgery required to squeeze his four books into one show. He became close friends with Andrews, and it was to her autobiography that I have turned for any trace of him in these latter years of his life. As it happens, there is a lot here. Tim clearly made a deep impression on her, and features prominently in the final chapter. She paints a candid picture of a by turns brilliant and difficult man, unpredictable when drinking, prickly when composing or writing. 'Get out,' he would mutter, without looking up, if disturbed at his desk.

OCTOBER

A chance has come for me to visit America again. The logistics of even this short trip 50 years on help to put those of White's epic tour in proper perspective. I land in New York City, which remains chaotic, frenetic, everyone talkin' loud (and apparently to themselves in many cases, in this era of hands-free phones). I have been invited to a symposium at Connecticut College, marking the 50th anniversary of Rachel Carson's *Silent Spring*, c/o her biographer Linda Lear. I take the train from New York along the New England coastline, distracted from re-reading Lear's weighty, meticulous and absorbing biography, *Witness for Nature*, by spectacular estuarine wetlands illuminated by high-definition sunshine beyond the window, cloaked in the dazzling colours of the Fall. Vast, rippling reedbeds are patrolled by harriers, shimmying over feathered stems in onshore breezes. There's an outside chance of seeing Goshawk. John James Audubon, the renowned early 19th century bird artist, knew the Gos from here:

"Along the Atlantic coast, this species follows the numerous flocks of ducks that are found there during autumn and winter, and greatly aids in the destruction of Mallards, Teals, Black Ducks, and other species ... He sweeps along the margins of the fields, through the woods, and by the edges of ponds and rivers, with such speed as to enable him to seize his prey by merely deviating a few yards from his course, assisting himself on such occasions by his long tail, which, like a rudder, he throws to the right or left, upwards or downwards, to check his progress, or enable him suddenly to alter his course. At times he passes like a meteor through the underwood, where he secures squirrels and hares with ease. Should a flock of wild pigeons pass him when on these predatory excursions, he immediately gives chase, soon overtakes them and, forcing his way into the very centre of the flock, scatters them in confusion, when you may see him emerging with a bird in his talons, and diving towards the depth of the forest to feed upon his victim."

Thursday

Connecticut College campus glitters in the October sunlight, the long sweep of its playing fields steering the eye towards the glittering Atlantic beyond

New London's harbour, a neat connection to the ocean, Rachel Carson's first love. A thrush I can't identify any more specifically is singing from the tip of a hemlock in the arboretum. A good omen. The biography has taught me that Carson once wrote to author and admirer E. B. White, "I can think of no lovelier memorial than the song of a thrush."

Friday

My base with friends Victoria and Richard in Brooklyn provides a chance to revisit Central Park on a hired bicycle, on my way to the Natural History Museum. Even before I've had to get off the bike and look I rediscover Pale Male, the resident Red-tailed Hawk. He is soaring proud over the boating lake, still in charge, monitoring the pigeons and squirrels below, and the passage migrants flitting through the park's canopies. He dips out of sight, and I carry on north to the reservoir. Here, the cries of a raptor make me crane skyward. Besides the wheeling gulls there is another Red-tail up there, a young one, with an entourage of 200 starlings following it; not mobbing, just sticking close, above and behind, enough to keep an eye on the hawk and perhaps to cause it some inconvenience. There are two more hawks higher up – adult birds. I follow them, getting in the way of the stream of joggers and power-walkers as I manoeuvre the bike to gain a vantage point. I get enough of a view to see the birds light on the northern pinnacle of a mountainous apartment block overlooking the Westside. I park up and sit down on the grass, against a low fence, to watch the hawks, with the aid of binoculars.

The tip of the building resembles an Olympic medal rostrum. One hawk is facing this way, crouched into the wind, leaning forwards, as though ready to launch at any moment. The other sits on the north side, partly out of view. They are on sentry duty, alert for passing rivals. Perhaps they might lead me to a visiting Goshawk. I watch for 20 minutes or more. The nearside hawk finally launches forth, and the other hops round to take its place. A little while later the second hawk pushes off and folds into a near-vertical attacking stoop. I lose it when it plunges beyond the canopy. Hurrying to the point where it disappeared, I find only pedestrians, runners, cyclists, children in a play area, migrant American Robins dismantling a berry-laden shrub, jittery, but otherwise going about their regular business.

The American Museum of Natural History is a few blocks south of here; a colossal edifice. My chances of finding a Goshawk here must be high, although even in death the bird has tended to evade me. I first find a bust in bronze of John Muir, the Scot who championed National Parks for America in the 19th century and of whom his Indian explorer friend once remarked: "Muir must be a witch to seek knowledge in such a place... and in such miserable weather." You can take the boy out of Scotland ...

I have less difficulty tracing stuffed American Gos. It is the first bird in the first glass case of birds on floor 3, Sanford Hall. This eastern subspecies of North American Goshawk is known to science as *Accipiter gentilis atricapillus*, which means dark-capped. It is distinctively different in appearance to European birds, not only in the blacker head and back, which make the blood-red eye all the more vivid and alluring, but in the flecking and subtle streaking on the breast, giving a greyer impression from a distance. Up close these feathers are beautifully and delicately marked, almost grouse-like. It isn't difficult to see how even the Victorians in Britain and Ireland could be sure when (if not exactly how) a bird from here had landed among them.

It is also helpful to see this museum exhibit for the physical, 3-D comparison it allows with the other raptor species around and beside it. It dwarfs the two other *Accipiter* species found in this continent. The smallest is the equivalent of our European Sparrowhawk. They call it the Sharp-shinned Hawk. It looks impossibly delicate and brittle beside the barrel-chested Gos. And there is an intermediate, which they call the Cooper's Hawk. Both smaller cousins are adapting to urban life much more readily than the Goshawk here, for reasons that would be fascinating – if difficult – to unpick. They are bantamweight and welterweight beside the middleweight Gos. If it isn't bossing Central Park now, perhaps it might do one day.

There is a more diverse medley of raptors and owls here than for European woodlands. The Gos has more competition, even though the ones that match the Gos in length – like the Red-tail – also appear slim-line alongside. The red eye and bold white stripe across the brows of the Gos give it its warrior-like appearance. In the southern Rocky Mountains of the US and Mexico lives the other, even larger and darker subspecies, even more facially striking, which they call *apache*. It shares this name with a type of pine that characterises these rugged ranges, and of course with one of the most renowned native American tribes of the west.

The importance of eagles in particular, and hawks in general, in indigenous culture is well established, but I have been wondering what significance, if any, the Goshawk might have had for the original inhabitants. On this same floor are galleries dedicated to Plains and Eastern Woodlands Indians.

I learn from these that for the forest-dwelling Indians hawks have been the 'swift flying birds', and carrying the skin of a hawk into battle confers the gift of speed and agility. There are feathers here too on a war pipe that I can identify as Goshawk. Some also believe that tribe members can become shamen, with supernatural powers. This gift comes in the form of dreams in which they are taught sacred songs and instructed in the use of medicines from materials like shells, bones, roots and animal skins. The shaman carries these in a bag, to cure illness and reveal hidden truths, and sometimes to conduct sorcery.

I wonder if a bird with the Goshawk's particular bravery and ferocity in defence of its nest would surely have garnered a particular fascination and respect from those sharing its domain, some if not all of the members of a given community. Perhaps any specific evidence for this would be unlikely to survive.

***

As it happens, while here in the States, I have received word from back home that the panel reviewing the US Goshawk that turned up in Scotland back in

1869 has now reached its verdict. The record has been rejected: not enough evidence that it got there under its own steam. There is also the possibility that the skin of the bird shot by gamekeeper Stewart, then taken to Brechin by someone else, bought there by someone else again and finally identified after being stuffed in Glasgow, might have been switched with another bird, perhaps imported as a skin from the US.

Might all the Irish records be similar fabrications or mix-ups? Could Goshawk skins be imported from the US and not carefully labelled as such, for collectors? Why pretend they were shot locally? Interesting to chew these questions over, and to conjecture on how the Victorians were thinking.

***

In the course of my explorations I find a vast multi-columned stone edifice of a building and realise that it's a Post Office on the scale of a palace, and climb the steps to buy a few postcards and stamps. The man at the counter is treating the line of young backpacker types with parade ground scorn. We are evidently not queuing to his satisfaction.

When it's my turn to have my chops busted I place my order for postcard stamps, then notice he has a sheet of raptor stamps on his desk. "Can I buy one of those?" I venture, pointing, braced.

"Those *aren't* the right kind of stamps for your cards," he barks.

"I understand that," I reply. "I'd just like one."

"Which one?"

"The Goshawk."

"You want the *Gasshaak*," he repeats. It's not a question. "OK. You want the one that's *right* in the *middle* of the sheet." This isn't a question either.

"Just to add to your problems," I tell him.

"I don't *have* any problems," he claims, adding "*Jeez*, when do *you* go home?"

I am wearing my sweetest smile.

I like the idea of this Goshawk getting across the Atlantic on a postcard, even for 85 cents more than I need to pay for postage. But I don't help its chances when I lick the back of it, not realising it is a self-adhesive stamp. When I go to post the card I've written, the stamp – the Gos – is gone. Post Office guy might have had the last laugh but luckily I find it in my rucksack, damp but still viable.

I catch up with Marie Winn and her documentary film-maker husband Alan Miller at their apartment on the upper Westside later in the evening, to find out the latest in the Red-tails story, and any sightings of Goshawk. Marie has still only seen that one bird here years ago. 'It caused quite a stir in the park,' she recalls. Both she and Alan have been avid T. H. White fans in the past, and I give Marie a copy of *America at Last*, of which I found another copy at the Strand second-hand book store earlier – among its eight miles of books.

Saturday

I journey with Victoria and Richard north along the Hudson River valley to upstate New York through the gold, green and russet tapestry of the Fall's forests, wisps of cloud clinging cobwebby to the hills, trees of every size and shape stretching across the landscape to vanishing horizons. There is more tree cover now than at any time since colonisation. The trees have grown back since much of the agricultural land has been abandoned. It's mostly secondary growth forest, but already it is maturing. Surprisingly, and in spite of this apparent abundance of sylvan habitat, the Gos is now officially a species of 'special concern' here in New York state, in part because of uncertainty about just how many there actually are. If the phantom hawk is tricky to count back home, imagine the challenges it poses here.

A few pioneers were able to find them. I'm reading some old records of Goshawk encounters here, compiled and put online by the Smithsonian. I love this: "None will dispute the Goshawk's title to a place among the Kings of Winter," writes Arthur Cleveland Bent, a century ago. "A big hawk ... broad-shouldered, compact, yet clean of build ... proud and resolute of mien, with brilliant orange eyes through which the fierce spirit of the fiery hearted warrior gleams at times like points of living flame; the Goshawk ranks second to none in martial beauty and in fearlessness."

I have searched the Internet for any other clues. An enquiry on a website called Native American Culture catches my eye. It is headed "A dead hawk", and is from a man who stops for a stricken hawk on the highway. "At first I thought the hawk was still alive," he explains. "Its tail and wing appeared to wave at me as I passed by. I drove back ... the hawk's body was still warm. For some reason I was compelled by the hawk. I felt as thought its spirit was calling to me, or perhaps a sign. I don't know."

On the same day he received a call at work confirming his retirement from the navy. He is asking for any spiritual insight into the effect the hawk has had on him. A man called Richard Sutton provides the expert's answer.

"Native people who are raised from childhood in the traditions of their ancestors may indeed find a special connection to hawks. Sometimes they are considered spirit guides through clan affiliation, or through specific hunting or seeking ceremonies. Usually such a person will dream of hawks regularly, and through these dreams begin to answer questions to what it is they are seeking."

Travel dislodges trapped memories and vivid dreams. I woke from one that stayed with me. There were birds in the sky – crows and gulls, maybe – milling around. I was with a group of people, like ramblers on a mountain, not people I knew. One of the birds stood out, and I thought it looked like a hawk. It was flying differently to the others. "It's a hawk!" I blurted out, and as it got nearer it became not a hawk, nor even a bird. It became an insect, a huge mantis, or airborne cricket. The strangers were scathing.

"Hawks can also be considered messengers that carry prayers," Richard goes on. "Their far-reaching sight is another attribute that is sought by traditional people when seeking a spirit guide.'

Victoria and Richard have a little house in the big woods, and we arrive there late in the evening. It is on the edge of the Catskills. US mountains often have these magical names, and I've known of these ones since reading a book called *My Side of the Mountain*, by Jean George, when I was still in primary school. It tells of a boy from the city who runs away to live in a giant hollow tree, and who tames a Peregrine Falcon. I am kinda living the dream now. And as though to confirm this, and the wildness of the place, a forest creature has been making home in the bathroom. In fact it has been trying single-handedly to remove an entire window frame, having entered through the loft. Its incisors have made short work of the inner metal gauze intended to exclude insects. From the pile of splinters and one or two droppings on the window ledge the chief suspect – and there are many – is Grey Squirrel. Cabin life in these wild lands already feels a bit like a siege.

The much larger animal scat I find next morning just 20 feet from the house can be traced to Black Bear. There is a critter called a Fisher living in one of the neighbour's outhouses. The closest European counterpart must be the Wolverine. White-tailed Deer move calmly away, as I pick my way through the local woods. I note what looks like an old hawk nest in the fork of a pine by the driveway. Perhaps the Barred Owl that Victoria has been seeing has had use of it.

I read up on the Goshawk's place in the scheme of things here. It seems they exist at lower densities than in Europe; that is, there are fewer of them, where they occur. They are also less productive, producing fewer young, on average. They are more restricted to forests, and less inclined to move out of there into more open country, or to nest in small woods and even more popu-lated places. They are more inclined to hunt mammalian prey. These differ-ences in Goshawk ecology, size and shape in the US probably reflect greater competition with other species, notably Great Horned Owls and Red-tailed Hawks, such as those now claiming dominion over Central Park. Goshawks in North America are likely confined to a more specific niche. There are other, closely-related hawk species there, more members of the raptor guild.

There is the Sharp-shinned Hawk, the Cooper's Hawk, and there are Broad-winged Hawks and others – what we'd call the buzzard or *Buteo* genus. In Europe only the Sparrowhawk among the *Accipiter* hawks is present.

Tuesday

After hanging out at Woodstock and sampling its lingering beatnik spirit, of which Tim White had been a little disdainful, we've returned to NYC. I explore Prospect Park in Brooklyn at dawn with a fellow Scot and ornithologist named Keir, now naturalised and making documentaries with Victoria. He knows of one previous occasion when a Goshawk has been through here. Today we make do with a dusky Merlin, perched high in a leafless tree surveying the meadows. The Merlin is another raptor species that has adapted to city life in America where they nest in trees. Back home many people are rooted in the idea of it as a ground nesting, strictly moorland bird.

My next destination is Ithaca, five hours by coach upstate, almost as far as the Great Lakes, to reunite with ornithologist and all-round handyman Nathaniel Hall Taylor. We met on board the restored rubber boom steamer *Ayapua* in the upper Peruvian Amazon two years ago, when he was studying a species of swallow. I thought of us as Butch and Sundance, Los Bandidos Yankees, on their South American getaway. We explored the channels and wetlands of this remote, pristine world. There were birds galore, including hawks of all shapes and sizes, waiting in the trees. We witnessed two attacks, on an egret and a cormorant, as we installed nest boxes along riverbanks. Our two-week adventure then and subsequent friendship has been punctuated by one-liners from the movie. "*Esto es en robo!*" is a particular favourite.

My first appointment here is to visit the forests of the Appalachian plateau with Matthew A. Young from the Cornell Laboratory of Ornithology. Matt has been part of teams studying the Goshawk here, although a couple of years have passed since he was last able to visit the nest sites. As we drive the wide valleys with their forested hillsides I am struck by its parallels with

north-east Scotland, albeit on a vastly magnified scale, and with much greater diversity of tree species. Even so, two trees well known to me and no more
native here than they are in the British Isles are represented – the Norway
Spruce and the European Larch.

The first forest we visit is selectively logged to promote ecological balance. There is no geometric clear-felling here. Matt's mentor has been a man
called Jim Spencer who has studied up to 25 or so nests spread across these
woods. Goshawks haven't been known in New York state for much longer
than in the UK, as it turns out. Jim discovered the first breeding hawks in
the 1960s, and at first he had difficulty persuading others that he had. The
colonising birds seemed to like the maturing plantations in particular. The
banding programme began about 30 years ago, but Jim is now in his eighties
and unable to access the sites the way he could until well into his seventies.
Although the breeding season is of course long past, Matt is happy to have
the excuse to get reacquainted with this little-visited forest we are in now,
and to introduce Nathaniel (Nat) and me to its secrets.

We find several of the old nest sites, our search helped in no small part
by the aluminium sheets that have been fastened round the trunks of the nest
trees to discourage Raccoons from climbing to rob the nests at night. I'm impressed by their intrepidity. From what I've heard it might have to be pretty
dark before a humble Raccoon could risk confronting even a sleepy female
Gos at her nest. Some of the nests have vanished, blown out by winter storms
or collapsed by weight of snowfall. But we find new nests, perhaps even this
year's. The birds are still around. Mind you, we have chatted so much and
so loud on the way in here that if we find any Gos today it won't be through
stealth. If it were spring, they would come and find us.

Matt reminisces about the hazards of approaching Gos nests when they
have eggs and young. He describes being chased unceremoniously out of a
forest by one irate female. It took him 20 minutes to get a hundred yards. "I'd
be running from one tree to the next, she'd be swooping at me, flying up into
a tree ahead of me and to the side, waiting for me. I'm trying to get to the

car, she's trying to get hold of me round the neck. I'm tellin' ya, these birds think they can take ya," he tells us, laughing a little nervously at the memory.

"Even when there's two of you, she still attacks. We found we needed three or four people, then she stayed away. But when one of you is climbing to the nest to band the young, you gotta wear layers of protection: helmet, face guard. And the guys on the ground are just watching – and glad it's not them up there."

***

The reputation of Goshawks for self-defence is long established. I've heard them say in America that you're not in true wilderness unless there's a chance you could end up as dinner. The following eyewitness accounts of scrapes with the Goshawk give some idea of what it must feel like to be prey. This one is from Clarence F. Stone, writing in 1921: "The female was on the nest, but when I was about 200 feet distant she swooped from the nest and attacked me in a most savage manner. I could have killed her but did not wish to. However, the attack became so ferocious, with lightning rapidity of swoop after swoop, that I was obliged to protect myself with a club ... From the time I strapped on climbers it was a question in my mind whether I could reach the nest with a whole scalp. Luckily the male goshawk did not appear, so I had but one angry and bold fighter to contend with. Before I reached the first limbs I was obliged to stall on my climbers and hug the tree with one arm while I flailed at the bird with a club. A score of times I missed but when she grabbed my cap and flew a hundred feet with it, I realized I must back down or else wing her.

"Back she came with speed of an arrow, wings half closed, eyes blazing, and uttering angry 'cacs', all of which meant that to save my scalp I must wing her. But she was so alert and quick that it was several minutes before I clipped her fore wing so that she fell to the ground, still full of fight and 'cac-ing' loudly."

According to Bent, an Albert A. Cross provided notes on "thrilling experiences" that he, Harry E. Woods and Lawrence J. Sykes had in Massachusetts. The group found a female Goshawk "in a very bad humor and hostile; she making four attacks on him while he was at the nest, coming at full speed and not uttering a sound. Woods was able to protect himself in a degree by pulling his coat over his head and dodging, but eventually the hawk lacerated him quite badly on the upper part of one hand and wrist."

Sykes eventually shot the bird after she attacked him on three separate days while he was fishing in a nearby brook. His story was covered in *The Springfield Union* of June 6, 1931, which reported that: "The bird caught the fisherman unawares the first time and tried to sink its talons, nearly an inch long, into his face and neck. He finally beat it off after it had circled and swooped at him a number of times. Not many days afterward, the fisherman went back to the brook again and the bird gave him another battle. The third set-to was the day before yesterday. The bird this time was more persistent than ever and in one of its vicious dives struck the fish pole and broke it in two. Yesterday the fisherman went to the stream with a gun. The hawk evidently saw him coming and met him some distance from the brook. The Springfield man who is an expert hunter as well as angler brought the bird down with two shots."

***

Dr. George M. Sutton reported in 1925 that he had been: "almost constantly attacked and screamed at by the female bird. Before my companions left me I crawled into a rudely constructed blind where I crouched motionless, hoping that I would not be detected by the hawks. The female bird drove the departing group of men to the edge of the woods and then returned, calmer for an instant or two, apparently, and then, spying me without the slightest difficulty, redoubled her fury and bore down upon me with savage intent. Intrepid and insistent she swooped at me from all directions and only the branches of the blind kept me from the direct blows of her feet although the protecting boughs cracked and snapped at each onslaught. My being alone doubtless increased her daring and she perched at a distance of only twelve feet and screamed in my face, her bright eyes glaring, and her powerful beak expectantly parted.

"... I photographed the attacking bird, and while I tried to steel my nerve to accept the blows of her feet without flinching, I found I could not. Every time, when I saw her glowing eyes, partly opened bill, and loosely poised feet descending upon me I ducked and raised my arms in spite of myself. Had I not worn a strong cap and a cloth about my neck no doubt her talons would have brought blood more than once; and it was evident that the claw of the hind toe was most powerful and effective, since that nail dug in and dragged as the bird passed on.

"The most memorable thing about the day's experience was the method of attack of the female bird, which has partly explained to me the ease with which some of these birds capture their prey. When the Goshawk left her perch to strike at me her set wings and slim body were for several seconds almost invisible and the only actual movement perceptible was the increase in the size of her body as she swiftly approached. Three times at least I was looking directly at the approaching bird and did not see her at all because the lines of her wings and body so completely harmonized with the surroundings, and the front view was comparatively so small."

I had heard about the bravery of Goshawks here in defence of their nests, but no idea the extent of their belligerence. What other animal would be so pre-emptively confrontational? This would be something to experience, which Matt has recommended for next spring. He can lend me a helmet and goggles, but I'm not sure I would wish to inconvenience the birds without a legitimate purpose. "Rare blood group too, mind," I have told him.

We have seen half a dozen nests today in trees of several different species, both conifer and broadleaf. In some studies, North American Goshawks have shown a preference for nesting in deciduous trees in mixed forests, even, it seems, when the deciduous trees are an isolated minority. One study in Alberta, Canada, around a century ago found 51 out of 62 nests in deciduous trees. In contrast to the aggression shown by some Goshawks, these birds appeared also to "require seclusion". This from another man named Young, no relation to Matt: "I watched my Newfoundland nest for nearly all of one day and parts of two others, but no hawk came near it. This may have been due to the presence of an observer, for I have noted that other hawks will not come near their eggs or young for hours if they know they are watched, and their eyes are exceedingly keen."

On the way back to Ithaca we stop by at a couple of houses where Matt has permission to check the garden bird feeding stations, to see what interesting migrants are being drawn in. We get chatting to one old fella who loves his birds – a Scott Campbell, proud of his Scots heritage – who, when he finds out about my research, tells me about his neighbour's daughter who is on the local soccer team and was attacked by a Goshawk while out jogging just down the road. "Formidable bird," he laughs. "I wouldn't wanna tangle with it." While we are chatting, a raptor looms up over the clearing that forms his garden, then veers away over his house when it sees us. Northern Harrier.

Thursday

While in NYC I had paused by the Passenger Pigeon in the Museum of Natural History. Stuffed birds have their place, but there is something profoundly arresting about being eye to eye with a species that is now extinct. I felt as though I should lay a wreath. But even that would be inadequate. Of all the birds extinguished by humanity the Passenger Pigeon's story must be among the most bewildering examples of our power. This was no small island endemic, isolated and vulnerable. This bird's population spanned the deciduous forests of half a continent, with its core population here in the east. Its flocks once darkened the skies above the woodland tribes and pioneer settlers on their westward expansion. It is said to have been one of the most numerous birds not just in North America, but in all the world. It was beautifully coloured and marked, and it could be a nuisance, and it ate a lot. No effort was spared in exterminating it. The last free-living individual was shot by a boy in Ohio in 1900. The last captive one died in a zoo on 1 September 1914, a looming anniversary that must be solemnly marked.

Its passing as a species must have had a substantial impact on the Goshawk here. According to Bent, in the days when Passenger Pigeons were abundant in Pennsylvania, for example, Goshawks were said to breed there commonly. By the turn of the 20th century the Gos had become "comparatively rare, and breeds only in some of the mountainous counties, where it can find extensive forests of mixed conifers and hardwoods.'

Sunday

I could not have come to this part of North America without visiting Hawk Mountain, on the Kitatinny Ridge of the Appalachian Mountain range that forms the eastern spine of the continent. It's another pilgrimage. It used to be a place where hunters gathered in the autumn to shoot raptors. A law was passed in this state in 1929 which put a $5 price on the head of the Goshawk. Men promptly went out and brought down raptors of all kinds. "Most men cannot, or will not, distinguish the good hawks from the bad," writes Bent. Then in 1934, thanks to funds put up by a donor called Rosalie Edge, Hawk Mountain was purchased and became a sanctuary.

I travel down with Nat and another ornithologist called Justin Hite, to witness the Fall migration for ourselves. Justin knows Goshawk from the western states. He recalls a time when the Yosemite National Park authorities removed a Gos nest after hikers had been attacked. "It wouldn't be dealt with like this today," he says, adding "but that is one bad-ass bird".

Seeing hawks here is guaranteed. In fact we will no doubt see loads, as it's peak season. Seeing a Goshawk remains an outside possibility. They tend to move south later, in November, as do Golden Eagles, although odd birds do pass through around now. On arrival, I check the sightings blackboard in the visitor centre. One Goshawk went through yesterday. The average for a whole season in recent years has been just short of 50. Early indications from other species are that there has been a general failure of tree seed crops in the northern forests, with birds of all kinds falling south in unusual numbers. We've already seen swarms of thrushes on the march. So perhaps there may be a greater than average surge of raptors to follow.

In preparation for the trip I've read an enlightening paper entitled 'The Periodic Invasions of Goshawks in North America'. It was published in *The Auk* in 1977, and describes Goshawk migrations each autumn since 1950. Goshawks seem not to be hard-wired to migrate, the way many other birds are. They will move in response to need, rather than because programmed to do so, to spend winter in a better-stocked, warmer place. So each year there may be no Goshawks recorded at a given watch point, or maybe just a few, or a trickle. In other years there is an influx. There were two dramatic Goshawk surges south in that time, the first in the '*Silent Spring* winter' of 1962 and 1963, and the second a decade later, 1972 and 1973 (incidentally, 1972 is the year DDT was banned in the US). This latter Goshawk exodus is thought to have been the largest ever recorded. It was so huge, and preceded such a spectacular population crash, that the authors (Mueller et al) predicted that it could be the last for several decades. The Goshawk population plummeted by as much as 70 per cent in its wake. It was thought that the population might take until the mid-1980s to recover.

It looks like the authors may have been right. Looking at the Hawk Mountain data on Gos numbers passing here over time, the mountain peaks of the graph show only foothills since that big year of the early 1970s.

Goshawks in the northern forests rely heavily on two prey species – the Snowshoe Hare and the Ruffed Grouse. When these undergo their periodic population crashes, the Goshawk has to look further afield. Competition between them will force less experienced ones out. While in most winters it is the young birds that have to move out, to find food, in these invasion years adult birds have been the ones displaced. Most passage birds of course head south, but some, it would seem, head east. And if you head far enough east, you hit the Atlantic ... this would be the likely source of Ireland's trans-Atlantic vagrants, if they are to be believed.

"Winter is the time when we look for the goshawks to swoop down upon us," Robie W. Tufts has written. He cites "well marked, heavy flights" recorded in 1863, 1889, 1895, 1898, 1905, 1915, 1916 and 1918; and particularly large numbers in the fall and winter of 1896, 1906, 1926, 1927.

The time has come to try to see one of these birds. We take our seats on the scattering of the generally rough-hewn but in places shoe-worn limestone rocks that provide the look-out points along the forested ridge. The gods are smiling on us and we have picked a perfect day for this, blue and breezy. Stray golden leaves are spiralling against the blue. What scattered tufts of cloud there are up there are speckling the miles-wide sweep of the valley with darkened patches. We pause at a few lookout points along the trail up. I pick up on a hawk circling low on the other side of the valley. Nat and Justin get their bins on it too. It isn't like any of the other raptors I've seen in the US so far. It is dark above, pale below, compact, purposeful, the right kind of size.

"Is it a Goshawk?" I'm wondering out loud as I trace its progress through bins. We work out as it reaches our side of the valley that it is a Red-shouldered Hawk – another of the many *Buteo* or buzzard species here. False alarm, but great to see one of these for the first time.

I suppose I have imagined solitude but there are hundreds of people gathered here too, of all ages and types, massing as though also on pilgrimage, or for a ball game, seated on the scatter-cushion rocks, soaking up the sun and the view. The main lookout point affords views to the north and west, across the high mountain pass with the valley sweeping to the south below, carpeted in a multitude of tree species, though dominated in now rust-coloured oaks, the boulder bed of the river visible in places in its midst. The ever-present Turkey Vultures swing this way and that, low over the canopy, wings held above the horizontal. I'm carrying with me a photograph of Rachel Carson taken in 1946 from one of her visits here. She is perched on a rock with the landscape of her home-state Pennsylvania stretching to the horizon behind her. Holding the monochrome photograph in front of the multicolour view we identify the very rock she sat on then. From her descriptions of the day it was colder back then, and her group had farther to climb to get here.

Like her, we sit back to enjoy the show. Raptors materialise like fellow pilgrims out of the north and eastern horizons. The game we are all playing is first to find them, then to speculate as to their identity as their outlines and forms slowly grow. Like any game it gets easier with practice.

"If we were Dutch hawk-trappers we'd have a shrike to help us," I joke. A little boy beside us gets really good at spotting them first. "Perhaps we don't need a shrike after all," I am able to add.

Word of the approaching birds ripples through the audience, and there are gasps and exclamations as raptors reach, wheel and dive over and past us. A semi-official man with a telescope on a tripod is calling the birds that he sees. "Sharp-shin goin' low!" is a regular shout from him, like an umpire calling a foul play. The Sharp-shinned Hawks are regular, passing like long-distance runners on a well-defined route below and sometimes level or higher than our eye-line. They are all in a hurry, beady-eyed, intent on their journeys, flicking short wings and raising banded tails, sometimes swooping with wings tucked, like little shuttlecocks, fixed on an earnest, centuries-old mission that must be annually repeated, regardless of who is watching, who is in on the secret. It is impossible to confuse a diminutive Sharpy, as they are known, with a Goshawk, even at a distance. Today produces hundreds of them, and we witness the passage of a high proportion. But there are also one or two large *Accipiters*, and moments of held breath, heart flutter and hints of Gos, but these turn out to be the mid-size, slower flapping Cooper's Hawk, a bird more apt than the Sharpy here to pause and wheel, to spiral high, less intent on getting south in a hurry.

Bald Eagles emerge from the clouds, national bird and crowd-pleaser. What I take to be a resident one rises from near the lake way below us on the plain, winding its way upwards against the far valley slope to engage passers-by in a dramatic dogfight, with Red-tails and Ravens weighing in.

Black Vultures slide by high in small groups. Red-tailed Hawks are regular and there are Merlins too, and one solitary American Kestrel. But so far no Goshawk. I find it hard to imagine the publicity-shy, loner Goshawk submitting to participate in so public a procession. It doesn't fit the image, somehow. I imagine that it would, if it could, make the southward march tree to tree, below the canopy, by some back door route.

By the time we leave, our ridge has thrown a shadow across the wide valley floor. There has been no Goshawk for us today, but as I explain to my comrades I am ok with this. I like that the bird is one step ahead of me. I don't want to own its soul.

The road home, like all roads here now, is lined with election placards. With two weeks to go, it's getting niggly. There are signs proclaiming "Don't frack with New York". We pass a fracking station on a distant plateau, with its gas flare candle burning bright against the night sky. There were none yet visible from Hawk Mountain, although we did pick out faint against the northern horizon a line of what must be giant wind turbines, harnessing the energy of the wind, just like the southbound hawks. If we can just find room for a few more of them, in places where they won't blight too many views or kill too many migrant birds, we might have a hope. The smell that follows us soon after for what seems like a mile or more is from road-kill Skunk. I'm glad when it stops, but I wouldn't describe it as unpleasant, in small doses.

Monday

"This bold brigand of the north woods, the largest, the handsomest and the most dreaded of the *Accipiter* tribe, swoops down in winter upon our poultry yards and game covers with deadly effect. He is cordially hated, and justly so, by the farmer and sportsman; and for his many sins he often pays the extreme penalty."
Arthur Cleveland Bent

In the morning, earning my keep, I help Nat assemble and raise a chicken barn roof at a neighbour's place, in a typically beautiful woodland clearing, with the oaks and Sugar Maples throwing leaves at us as we hammer and saw in the sun, and the chickens and guinea fowl scurry and fuss around us. The little white silky bantam with the Marge Simpson bouffant and furry slippers looks especially unsustainable in the context of the forest beyond, as implausible as a handbag poodle in a Wolf pack. I befriend an impossibly cute chipmunk and play peek-a-boo with it at the base of a hemlock. In a forest full of dangers it may secretly like having us – and the chooks – for company.

Dr. George M. Sutton spent a whole day in 1925 watching a Goshawk nest. He found that "the young had evidently been fed almost altogether on chipmunks – although fur and some small bones of gray and black squirrels, weasels and white-footed mice were also found".

At one stage the motley farmyard fowls assemble under a bush, pressed close together, making an almighty row. I scan the blue yonder and within a few seconds an ivory-chested hawk is cruising across the clearing. Red-tail, as usual. The owners here report losing a chicken recently, finding only its plucked remains. Like most folks hereabouts they accept this as an inevitable tax on living in the forest. Replacement chickens are not in short supply.

I dust myself off after lunch and Matt leads me on a tour of the Cornell Lab, with the vast Macaulay Library of Natural Sounds in which he works. In the elegantly panelled meeting and lecture hall there is a collection of original artworks by Louis Agassiz Fuertes, and my eye is drawn immediately to his painting of a Goshawk capturing a Spruce Grouse. We drop in on the collection of bird skins, a chill, windowless vault with row upon row of tall metal cabinets. Archivist Tom Schulenberg leads us to the doors marked with the names of the bird of prey families. We find 'Accipiters', although I note that *gentilis* isn't among those labelled. I wonder if I may be poised to fail once again to find the bird. I needn't worry.

Tom twists the handles like a safe opener, and pulls open the double doors to a pungent gust of preservative, to reveal wide trays, also neatly labelled. I am now looking at no fewer than four drawers of *Accipiter gentilis*, each containing eight or so stuffed specimens, lying side by side, and beak to tail, like a row of very stout clubs. This impression is heightened when Tom slides one of the trays towards us and lifts out one of the stiff hawks to allow a closer look. It is eyeless of course, the iris livid-red in life now replaced by a white orb, giving a ghoulish feel to the otherwise elegant corpse.

Hand-written, yellowing toe-tags on the clenched talons indicate the origin of each specimen. The ones we have time to examine most closely are from Pennsylvania, in the main, some from the 1930s. I wonder if they were shot at or trapped at the Hawk Mountain site, before protection there in 1934. There is also a much darker bird, an example of the *apache* subspecies, from the western state of Nevada. These are mostly adult birds, and look in some cases huge, which Tom explains may be partly because of stretching that can occur during the preservation process. I am struck by how little size difference there is between the sexes. Perhaps the male here needs to be big too, given the large prey and many adversaries he might have to tackle, not to mention the hard winters he is expected to endure.

***

I hadn't expected to find a Pheasant-rearing pen amongst the many hundreds of miles of native forest here, but when I hear that there is one near Turkey Hill, not far from where I am staying, I ask to be taken there. I say pen, but in true American style this place is vast. It can be measured in miles rather than metres. It must be half a mile across, and several long. Most of its compartments are crammed full of Ring-necked Pheasants, just like the ones we have in the UK, males segregated from females. This government facility rears tens of thousands of birds for shoots in a number of release sites. It is the last such place in the state. Pheasants have naturalised in small numbers around fallow cropping areas in the west of the state, but have declined as woodland has regenerated. Even if they survive the guns, few released birds survive the winter. Wild Turkeys, meanwhile, have increased, along with deer and Black Bears.

There has historically been less illegal persecution of the Goshawk in North America than back home in Europe. A study by Sperber, in Glutz et al, in 1971 found that almost half of all Goshawk nests in Europe were failing at that time. Of these, 84 per cent were disrupted by humans. Two-thirds of Goshawks were dying in their first year. A third of older birds were also dying annually. In North America, Goshawks tend to be further away from people, and harder to find in the wilder forests. Hunters here tend to respect raptors and their right to a share. America also lacks people paid to manage game birds, pigeon fanciers and farmers with domestic fowl in open pens.

Another Red-tailed Hawk cruises casually overhead against the azure sky. I can see several Turkey Vultures black and wobbly in the near distance. Nat tells me that some days there are dozens of *Buteo* hawks in the air here, waiting for scraps or escapes. But there are no crow traps and definitely no guns poised around this farm. Americans don't as a rule kill hawks. These Pheasants are safe while confined, because despite its vast scale, this entire rearing facility is covered by a mesh net roof, supported at regular points on clothes pole-like props. Of course it is. It would be unfair to the occupants, as well as careless, not to roof it.

***

The last word is reserved for Herbert Ravenel Sass, who wrote in 1930:

"We do not live by bread alone. Beauty and courage, swiftness and strength mean something to us; and we shall find these qualities in high degree in the hawks of the *Accipiter* elan. Especially is this true of the largest and strongest of them, the Goshawk, one of the deadliest, handsomest, bravest birds of prey in the world."

Chapter 13

WE NEED TO TALK ABOUT THE GOSHAWK

Monday

First thing in the morning, I am at my desk, opening files, firing up the machine. I turn to find a colleague (I'll call him Don) standing over me, clutching a large mug in one hand, and tapping with the palm of his other hand. He lives in a village not too far from mine, on the edge of small woods and parkland, the ones where I thought I saw that male Goshawk three years ago, the day I also found prey remains in the cemetery, the dead crow with the apparent injuries, the squawking woodpecker. Since then there has also been some hearsay from a retired gamekeeper. In fact we were talking about it again only last week, Don and I, when I was querying the "huge Sparrowhawk" he'd seen there a few months ago. He's always maintained that although he himself has never seen a Goshawk there, his mind is open that they might be present. This morning he has a look on his face that I can read, even before he has spoken. It's the unmistakeable air of a man who's made a breakthrough.

"What have you seen?" I ask him. "It's Goshawk, isn't it?" I add, as a smile breaks over his face. He is nodding, slowly, eyebrows raised, like a man who has seen an apparition, or at least had an epiphany. "Gos*hawks*!" he whispers quite loudly, correcting me slightly. "Absolutely amazing!"

We are both laughing now. This is how all working weeks should begin. He goes on to explain how yesterday morning he had noticed these raptors from his garden. Although they were some way off he was able to get his telescope onto them, and found that they were displaying.

"At first I thought they were harriers," he recounts, excited. "Long tails, stiff wings held upwards at an angle, white rumps, but then I could see the white was visible underneath too, there were deep wing-beats and then one plummeted vertically downwards and straight back up again. No doubt at all these were Goshawks – displaying. I had no idea they displayed in autumn."

He says he has checked with someone else in the village, another ornithologist, who now reveals to Don that he has been seeing Goshawks on and off locally too, for a while. He stopped reporting them because he has never been believed. He's tired of being accused of what is disparagingly known as 'stringing' – over-claiming, in other words.

***

I'm catching a mid-afternoon train to London, for an event at the Royal Botanic Gardens at Kew. The sky is clear and a cool breeze blows from the north-east. I scan the sky from the platform, now more than ever alert for the remote but real possibility of Goshawk up there. I notice a falcon I might otherwise have missed, and as it disappears from view I hurry to the steps to get on it again from the railway bridge. From this vantage point I am treated to an extraordinary spectacle. The Hobby has been joined in the sky by a Jackdaw. At first I assume that mobbing must ensue, but in the several minutes of aerial ballet that follow not once does the corvid attempt to strike the falcon, and not once does the falcon feel the need to take evasive action or attempt to shake off the Jackdaw. Instead, they fly alongside one another, a few feet between them for the most part, the swivel-hipped master Hobby apparently intent on hawking insects, the apprentice Jackdaw engaged in an at times hilarious break-dance of twists, spins, flips, flaps, dives and clambers-back-up.

It is as though it is trying to mimic, or emulate, or mock its partner in this crazy waltz, or even to send itself up. The pair are evenly matched in size, but the contrast in their movements could not be more stark – the Jackdaw all fits and starts, angles and jitteriness, the falcon smooth, languid, all gentle curves, sinewy effortlessness, almost ignoring its taunter as it concentrates on the feeding task in hand. I can only liken it to a dance-off between John Travolta and Ricky Gervais, a show-down provoked by the latter's famous comic take-off of *Saturday Night Fever*.

The Jackdaw eventually tires and gives up, as though snapping out of a trance, and flaps off towards town. Two more Jackdaws pass by moments later on the same route, and show no interest in the Hobby, which is still up there, carrying on its solo ballet, when the train pulls up. I scribble notes on what I've just witnessed. I wonder what the others at the platform might have made of me, staring up at the sky, if they even noticed. I had hoped in a way that someone might have asked me, so I could have shared the pleasure in this spectacle with them.

What was the Jackdaw doing? I think it was answering some conflicted combination of instincts – to mob the menacing-looking raptor, and – finding it not actually that threatening – and maybe also finding it in a sense 'inspirational', as I do, deciding instead to display its own fitness, its own repertoire of stunts. The tendency of corvids to playfulness, as well as mimicry/emulation is well established, and perhaps here was an example of it from the wild, across a species boundary.

Thursday

A day off. I wake to clear blue skies and even a whiff of frost. I head over early to Don's village to have a look for myself, revisiting a place I haven't actually been to since the spring, having lost some of my belief that the birds could actually be present or that I would ever find them. Now, my enthusiasm for the place is re-fired, and I have a glorious morning with the sun on my face, scanning the skies, watching the busy traffic of birds moving around against the autumn blue, following up on all the complaints of the corvids, the Chaffinches, the passing larks and thrushes. I even try a few rudimentary whistles but succeed only in vexing a nearby Robin and, I think, unsettling a Woodpigeon, which shifts in its seat high in a beech to my left. I scour the cemetery again where I found the prey remains three autumns ago, and pause at the spot where the putative Goshawk shot over our heads that otherwise sombre, drizzly afternoon. And now I dare to believe that we had been right. I feel vindicated – sane, even – daring also to hope that the Goshawks have been here all the while, breeding, settled, secure. As I pause beside some magnificent, mature Plane trees, chestnuts and even an elm, a raptor goes up from the line of pines on the opposite side of the field. Buzzard. I keep my eye on it. As ever, it might be enough to bring in other raptors. It duly does. Two more go up, both clearly Buzzards too.

A small falcon appears, and as it draws closer I work out that it's a juvenile Hobby, still here, not yet drawn south for an African winter. House Martins are passing over high, in flocks of a dozen or more. I put up a Kestrel from the tall hedge, and it flickers across the rough pasture, calling, as a Jackdaw tries to intercept it. Jays do the butterfly stroke across this open space too, and a Magpie passes over, higher, both looking desperately slow and exposed as they go, as though not yet aware of the risk they might be taking by leaving the security of the trees. And I hear again the yaffle of a Green Woodpecker, same place as before, only this time not sounding panicked or hysterical.

I sit on a bench in the cemetery as the mounting warmth allows some layers to be shed. I reflect on the Goshawks I thought I saw in spring of last year, a few miles to the north of here, the female soaring and then setting forth on a direct course right towards these woods, the male going into attack mode at the Jackdaws and Rooks milling around over the paddocks there. I am now quite sure of what I saw that day, and think it likely the pair were on a hunting and territory confirming expedition from this base. I permit myself a quiet glow of vindication.

Saturday

It has rained for much of the night, but I have woken to chill blue skies. There is mist across the road, and a ribbon of puddles as I ease the car round for the short drive to meet Don. "Is this the raptor camp?" I call over the garden gate. He lets me through, laughing, and goes to get the cups of tea while I carry his scope to the end of the garden, where he has a viewing platform, as though purpose-built for this very occasion.

"I can't believe after all these years your special Goshawk viewing tower has finally paid for itself," I laugh. "Yes, it's all been worthwhile!" he grins, carrying both mugs of tea in one hand as he climbs up the greasy wooden rungs of the ladder to join me.

We man the sentry post for an hour or so, on a morning alive with birds on the move again, and some, like the dozen Grey Partridges, static in the sun's warming rays on the edge of the ploughed field in front of us, with its wide margin around the perimeter. Don points to the place above the distant fringe of woods where he picked up the Goshawks in his binoculars while walking around the field, before racing back to the platform, and the scope, to get a better from here as the birds circled and displayed.

Today there are buntings, finches, pipits, the odd warbler still. A flock of ten Jays wafts past from behind us, to negotiate the wide open space of the bare earth field. They labour in the cool still air, working hard like salmon against rapids, or ducklings on a spate, looking vulnerable, clumsy, especially the one that has lost its tail in some recent skirmish. Another eight follow in their wake, in a loose group. They have been arriving from Scandinavia in unusual numbers. Swallows and martins are still moving through, and the Hobby races low over the far end of the field. Corvids rise and circle, and each time they do we follow them, to see if anything else is among them, anything causing them unease. But for some reason – perhaps the cold, the lack of air currents yet – there are no raptors up this early morning, not even Buzzards.

We decide to walk in that direction, leaving the scope behind. A Little Owl is sunbathing on a shed. Late butterflies investigate scabious flowers in the field edge. Yellowhammers flit past in groups. As we are walking up a lane there is a sudden, tremendous commotion among the ducks and Moorhens on a small pond almost completely obscured from our view by thick hedgerow vegetation. There is much splashing and panicked squawking, and pigeons scattering from the treetops, followed by other pigeons coming over to have a nosy. I find a gap and crane to look. I can see nothing but unsettled wildfowl. Don is beginning to get what this Goshawk business is all about. "It's like Lions in the Serengeti – you find them by watching the behaviour of the other animals ..." is his analogy. I think of the Goshawk more as the Leopard or, in a UK context, the Wildcat.

The morning warms steadily, the trees are now drenched in autumn sunshine, the wet wood steams in places. A huge regiment of mainly black-backed gulls clusters in the centre of another wide field of bare soil, studded with overlooked onions. A Buzzard stands near the margin at the far end, picking up worms. From this edge of the village we have a vantage point to scan the woodlands lining the horizon to the west, with the turret of a church the only thing breaking its even profile. By now there is a constant procession of Jackdaws and Rooks and even a pair of Ravens gets up, to soar on wide wings, dwarfing the other corvids. A lone Buzzard has now passed us, low, arriving late for the air show, and landing on the stag's antlers of an oak. A Sparrowhawk beats along at treetop height towards and past it, with barely a nod. Don has picked something up in his binoculars. "What's this now?" he mutters. "It's moving fast ... it's going like a bomb!"

I try to follow the line of his gaze. I can't get on it at first, but after it has evidently pulled out of its long, diagonal dash above the distant tree line, I settle on the shape of the bird in question. It is gaining height, distant and turning in familiar tight, quick circles, beating occasionally, tipping on flat wings like a fighter plane. As it banks, it glints pale on a wide body. It has all the hallmarks of Gos. Nothing else is near it. It stays up there for what must be several minutes, disappearing from view I think when in profile, then reappearing. It begins to make its way steadily back in the direction from which it came.

My arms are aching and wobbling a little from the effort and tension of staying with it. We both lose it at the same time. It vaporises. Maybe it folded and dived back into the wood. Another hawk appears from that area – more clearly a small hawk, a Sparrowhawk. We have worked this out even before it has begun to head towards us. It makes a half-hearted dive at a Goldfinch, which easily evades it after a brief pirouette by the pair. It lifts again and circles above the gulls, irritating but not panicking them.

Don and I discuss what we've seen, and conclude that the bird before it was almost certainly a Goshawk – perhaps just a little too far away to claim one hundred per cent. Following his views of the displaying pair last weekend, we think it's pretty conclusive.

I wonder whether Don would have looked properly a second time at these distant raptors in the sky if we hadn't had those earlier conversations about the possibility of Goshawks here. He clearly has the bug now. "I'm really keen to work out what's going on there with these birds," he says. "You can waste a lot of time looking for Goshawks," I tell him. He nods, and smiles. I think he understands what he might be letting himself in for.

We have agreed that the news should not be broadcast and that for now we should think about how to play it with the local landowner. There is Pheasant-rearing and shooting on the estate. Periodic shots ring out, to reinforce the point. But I have reason to believe it's a low-intensity, sustainable shoot, with none of the vast, unroofed Pheasant pens that characterise so many shoots elsewhere. It is known that the latest lord of the manor is aware and proud of the estate's recent record of maintaining recovered species like the Buzzard. But does he know about the Goshawks, and are the birds in fact breeding, or now trying to settle to breed? Can we hope and trust that the Goshawk is being welcomed and nurtured too?

I leave Don to get on with his life, and I have another quick search on my own, from another angle on the woodland, from some adjoining parkland with scattered mature oaks and other specimen trees. I can hear a bell, like a cat's bell, jingling on and off. It takes me a few minutes to realise that it might be a hawk's bell, as it moves from one of the trees apparently to another, and is accompanied by a tumult of Jackdaws, spilling out of the crown of the trees, and spiralling around over and among them in a seething, sneezing mass. Three Buzzards rise, languid, slow turning circles, relaxed. The crows spin and fuss among them, the ever-present noisy neighbours.

Further on a Red Kite turns up over the trees, then another, then a third. More Buzzards are up, and bleating. It is warm enough for them all now to spread their wings.

I call Sara to share the good news. "We've found the Goshawk," I tell her. "Nearly definitely ..."

After lunch I return. The day is so glorious I have to maintain the vigil, from the edge of an open field on the other side of the ridge, also an elevated spot, with woodland forming an almost complete circle on the horizons around me. I sit on the margin between two crops, back resting against a sapling in a Perspex tube. I set up my telescope on its old metal tripod on brass spikes, and scribble notes. When I point the scope at apparently empty skies I immediately pick up a cavalcade of soaring birds – all those milling Jackdaws again, like distant mayflies, Buzzards wheeling large among them at regular intervals. The sky beyond produces its usual steady procession of airliners coming in and out of London's northern airports, and vintage aircraft from local airfields of the keen amateur aviators. It is a day to be airborne now, clearly. The air show takes place against the soundtrack of a wedding at the church on the hill, and a clamour of bells fills the landscape; but there is no Goshawk this afternoon, within the pageant, that I can be sure of. I peer closely at a pair of raptors soaring in the general vicinity of the bird we saw earlier, but they are too far away to identify with any certainty. Driving past the precise location of Don's displaying birds from last week I find now freshly plucked Woodpigeon feathers by the road, some in small clumps, moving in the light breeze.

Although I don't find it again today I am luxuriating in the knowledge that the Goshawk is in there, somewhere. The landscape has taken on a new richness. I love the idea that the Goshawk makes the occasional appearance on the sky stage above these woods, that the other birds are having to make room and allowances for it, that it is the source of regular agitation, and a magnet for their attention. The woods are a different place now. It may be centuries since Goshawks could breed unmolested in this part of the world; at least two centuries since they could breed at all. It feels as though a natural order of things may be restored by the return of the exile, the prodigal bird, once feted and worshipped, then cast out, banished. What I love as well is that I think I'm finally sure what they look like now. I recognise the Goshawk, and I can begin to see the recognition dawning on others.

On the way home I drop in at the farm shop, which specialises in local produce, including game from the estate (could they market this as 'Goshawk friendly' one day?) and even Rabbit, making good use of the abundance of these on the farm's hundred acres of pasture. Not for them the discarding of Pheasant and other game corpses that I've heard goes on elsewhere.

I'm not against shooting, although killing things wouldn't be my or the vast majority of the population's idea of recreation. In fact it mystifies me, in the context of our artificial landscape, and I regret that it makes wildlife more fearful of humanity than it would otherwise be, and therefore harder to approach. It also makes the countryside in many cases a less welcoming and accessible place, and contributes to many people's continued disengagement from the outdoors. But I am told that without shooting, many estates would have no financial incentive to maintain even the scraps of woodland that have survived publicly funded agriculture, and the landscape and many wild birds would suffer even more habitat loss as a result. But I do think that unless we can rely on more estates to follow the progressive, sustainable model that I fervently hope has been established by this one locally, then we have to look hard at measures to help them get there.

These might include greater emphasis on properly securing Pheasant pens, reducing the size of these and the number of birds that should be reared in them, reducing the perceived need for the 'conveyor belt' killing of animals like Stoats and Weasels, which, for reasons that seem anomalous, enjoy none of the protection afforded to their close, mustelid relatives, the Otter and the Badger. The killing of Foxes too, which for conservation reasons may sometimes be an unpleasant necessity, the general, open-season destruction of these animals strikes me as frequently gratuitous and ineffective. It's the 'sump effect' again – shooting a Fox at a 'honeypot' creates vacancies for other Foxes, which in turn are shot or trapped.

"Keepers create hotspots and kill the birds that are attracted. It's like shooting Siskins off a peanut feeder," Mick has told me.

"The most vulnerable bird of prey species, and the ones that became extinct in the UK – Harriers, Kite, Goshawk, White-tailed eagle – are those that wander widely, have large overlapping home ranges, and congregate in numbers in hotspot foraging places."

It's about basic decency, in the end, and restraint. Treating nature with respect. The return – or otherwise – of the Goshawk to our lives strikes me as being an acid test of how decent we have become.

When I get back home from Don's I wander down the garden. I find a scattering of pigeon feathers beneath the Bird Cherry, dusted on the damp leaves it has shed. A few metres away are tail feathers. They weren't there first thing this morning. The timing is spooky.

The discovery and confirmation of the Goshawk so close to home has convinced me that there have been Goshawks in my village at least in passing during the course of my four-year search, since my encounter with the dead-yet-alive bird in the junk shop.

My local wood is just a short hunting flight away from Don's, and I am sure it would fall within the foraging territory of any birds based here, as well as being the kind of enclave that wandering birds would enter and even shelter in for a period, within a landscape where such suitable places are widely scattered. Looking back now, I feel confident that at least a few of those hints and glimpses have been of the real thing. I have no doubt that the mystery hawks in fenland, 20 miles downriver from here, were the real deal too, even if they have now gone again.

I've sent Mick a copy of *Silent Spring Revisited*, as a token of my gratitude for all his expert advice, so generously given. The book seems to have affected him, perhaps, stirring some suppressed regrets. "I much enjoyed the book, it brought back memories of the past – music and news events, politics and literature, colleagues and friends... As a new graduate, I held the view that better information, greater understanding of the natural world was the key to conservation progress. Well ... that is true, but information and understanding is persistently 'trumped' by wealth and power. It rapidly became clear that many conservation battles were lost despite the best of ecological research done."

I am pleased that the book's upbeat ending has helped. "I have a similar tale with a Goshawk nesting 80 feet up in a Sitka," Mick recalls. "The nest tree is ten paces from a main road, there is a public footpath below it where dogs walk every day – but the bird seems oblivious to it all. I use such little bits of good news (Speckled Wood butterflies have colonised locally in the last four years) to keep going, but your book obliges me to stand back and 'take stock'. So much of the richness of the countryside is gone. The current generation of child naturalists will not be able to experience the world I lived in."

And finally, a report on the breeding season gone by. "Not so bad a year as we thought when chicks were dying in June," he says.

Monday

I visit Scotland again at new year. On the road north to a cottage in Wester Ross, I stop off again at Perth Museum, to try once again to see the Goshawks in the vault. When I get there, the gate across the front entrance is closed.

Friday

I try again on the road south, from the treeless north-west, with its restored White-tailed Eagles, to the forested valleys of the Spey and Tay, from which these last Goshawks were taken. This time I am in luck. Curator Mark Simmons is in, and he takes me down to where the birds are now stored. Here at last is the pair, frozen in time. Toe-tags record them shot not in 1883, as is often stated (in fact that is the year they were procured by Millais), but in June 1879. They were killed by a Mr J. Boath, his own minor place in history assured as a result. They are mounted, with bits of scenery, and were removed from public display some years ago, perhaps because they had been bleached to ghostly forms by so many years in the spotlight. But this doesn't really explain the absence of the Goshawk from the public displays upstairs. There is another specimen here, a magnificent female, on another shelf, mounted on a branch, origin unknown, wearing a polythene wrapper. Hidden away. Perhaps, in an odd way, this continued anonymity is fitting. But maybe now the wraps should be taken off the Goshawk. We should talk about it more.

Whether we recognise it or not, the Goshawk was here for millennia before us, and may be here again after we have gone, unless we take it – and most of the other large and some not so large life forms – with us. Unless something changes, and we learn to revere the natural environment that supports us, our brief period of domination will end. We are in a sense a temporary, occupying army in their land, the Goshawk one of the guerrillas in our midst. Their struggle reflects back at us something of our folly.

As for the corner of the forest I might just get to share with them, I have worked out that the Goshawk is here after all, but that till now it comes and goes. That it used to breed elsewhere in my county is a matter of record, as is the fact that it subsequently stopped breeding here. Why it stopped, and doesn't now, or never has, in so many other places, is a matter of conjecture, and each person must make up their own mind, where direct evidence of interference is missing.

But I hope that we will shake off the illusion that this is a bird for which we don't have the right habitat. Mature woodlands have the structure, open fields have the food. Inner cities, it would seem, have both; may be perfect, even, if the Goshawk can ever get there. The Goshawk is coming back, and has been trying to throughout my lifetime. It's time now that it was permitted to find room again in our lives, for us to work with the gods' havoc, instead of fighting it.

Looking for the Goshawk has given me a fresh perspective on landscapes and woodland interiors, and helped me notice a lot of things I would otherwise have missed. When the spring comes again with any luck I'll still be looking, and the search will get easier, and even more rewarding. I might find the Goshawk, but it would be even better if the day could come when the Goshawk were allowed to find us.

Postscript

I've just learned from Japanese Goshawk man Takashi Kurosawa that the Goshawk has occupied Tokyo, with 20 pairs in the city.

Nat reports from Ithaca that the roof of the chicken shed is holding up well, but that, sadly, Marge Simpson is missing in action.

And closer to home, the Goshawk has also come back to the Queen Elizabeth Forest Park, in the heart of Scotland, where I began this tale, on that afternoon searching the forest with Dave, and Hamish the terrier. There are two pairs of Goshawks there now, and they've raised seven young this year. Dave has even taken a photo of one of these young birds. It was caught on his camera trap, glaring at the lens, as though cast in bronze.

Let's hope they are still out there – somewhere.

Epilogue

And maun I still on Menie doat,
And bear the scorn that's in her e'e!
For it's jet, jet black an' it's like a hawk,
An' it winna let a body be!

Robert Burns
Composed in Spring
1786
(A song sung to the tune of *Jockey's Grey Breeks*.)

Acknowledgements

"Writing is a lonely occupation," Rachel Carson once said.
"Of course there are stimulating and even happy associations with friends
and colleagues, but during the actual work of creation the writer cuts
himself off from all others and confronts his subject alone.
He moves into a realm where he has never been before –
perhaps where no one has ever been. It is a lonely place …"

I'd like to thank the following for their company in this search:

Umberto Albarella, Brigid Allen, Duncan Allen, Elizabeth Allen, Rainer
Altenkamp, David Anderson, Robert A. Askins, John Atkin, Steve Bale,
Dawn Balmer, Michael Berington, Rob Bijlsma, Keith Bildstein, Caroline
Blackie, Steve Blain, Judith Bond, Ariel Brunner, Peter Buchanan, John
Busby, Charlie Butt, James Cadbury, Peter Carroll, Clare Chadderton,
Jacquie Clark, Tim Cleeves, Marcus Coates, Mark Cocker, Paola Coccozza,
Stephen Coleman, Martin Collinson, Juergen Dammgen, Ian Dawson,
Malcolm Dennis. Dave Dick, Paul Donald, Brendan Dunlop, Mark Eaton,
Jose Edrich, Aniol Esteban. Brian Etheridge, Rhian Evans, Sara Evans,
Mike Everett, John Fanshawe, Andre Farrar, Marc Fasol, Viktar Fenchuk,
Nigel Fletcher, Lee Fuller, Graeme Gibson, Graham Goodall, Maxine
Gordon, Stewart Goshawk, Emma Griffiths, Tim Grout-Smith, Sheena
Harvey, Malcolm Henderson, Richard Hines, Justin Hite, Ben Hoare,
Angus Hogg, Mark Holling, Deirdre Hume, Rob Hume, Robert Hume,
Harry Hussey, Digger Jackson, Richard James, Bill James, Kay Jameson,
Kevin Jameson, Shelley Jofre, Sarah Keer-Keer, Norbert Kenntner, David
Kent, Andrew Kitchener, Jeff Knott, Marcus Kohler, Michael Krause,
Reiko Kurosawa, Takashi Kurosawa, Lars Lachmann, Robert Lambert,
Clark Lawrence, Paul Leafe, Linda Lear, Tasso Leventis, Christopher Lever,
Jeremy Lindsell, Brian Little, Toni Llobet, Jane Logan, Paul Marten, Helen
McDonald, Bob McGowan, Duncan McNiven, Mick Marquiss, Eric Meek,
Jon Megginson, Tim Melling, Matt Merritt, Hugh Miles, Alan Miller,

Patrick Minne, Dominic Mitchell, Isabel Moorhead, Allison Moorhead, Ian Newton, Piers Nicholson (and the website www.maxnicholson.com), Barry Nightingale, Peter Newbery (who has very sadly passed away), Martin Oake, Darren Oakley-Martin, Sarah Oppenheimer, Naomi Oreskes, John O'Sullivan, Ciril Ostroznik, Daniel Owen, Benjamin Panciera, Jemima Parry Jones, David Payne, Nick Pears, David Pennington, Julia Pennington, Steve Petty, Giovanna Pisano, Bernard Pleasance, Richard Porter, Caroline Pridham, Clive Pullan, Frank Pullan, Keir Randall, Nigel Redman, Sarah Richards, Roger Riddington, Stroma Riungu, James Robinson, Chris Rollie, Marc Ruddock, Christian Rutz, Steve Sankey, Norbert Schaffer, Tom Schulenberg, Guy Scott, Mark Simmons, Robin Standring, Lucy Stenbeck, Patrick Stirling-Aird, Marianne Taylor, Mick Taylor, Nathaniel Hall Taylor, Lisa Thomas, Andrew Thorpe, Ian Todd, Mike Toms, Gus Toth, Richard Toth, Victoria Toth, Roger Upton, Zoltan Waliczky, Jamie Wells, Steve Wheeler, Pat Whitfield, Joane Whitmore, Alasdair Wilson, Marie Winn, Melissa Worman, Derek Yalden (who sadly died in February 2013), John Young, Matthew A. Young.

Thanks also to my colleagues, in the Investigations team and beyond, working with the Police Wildlife Liaison Officer network. And to the FDT cats – for the t-shirt, and the moral support ...

I received a Roger Deakin Award from the Author's Foundation (Society of Authors), for which I am extremely grateful.

I am greatly indebted to Steve Wheeler, Jose Edrich and famil for their kindness and to Professor Ian Newton, for expertise so warmly shared.

Thanks too to the Rare Breeding Birds Panel (RBBP), to David Higham Associates for granting generously permission to quote from the works of T. H. White, to Piers Nicholson, for permission to quote from *Birds in England*, by Max Nicholson – *www.maxnicholson.com* and to David Hosking for permission to quote from Eric Hosking's *An Eye for a Bird*.

And most of all thanks to Sara, for helping me to look, and to see; even when you couldn't find your glasses.

Historic records of Goshawk in the British Isles

BEDFORDSHIRE 1950 ~ The first mention of the species in the county was of an escape in the summer of 1950 but the exact whereabouts was not disclosed. *Dazley and Trodd*

CAMBRIDGESHIRE 1950s ~ Curiously, there are nine records widely spread across this decade, and none from earlier decades, although Evans and Lack mention a young bird obtained in the county which is in Saffron Walden Museum. *Bircham* • **1970s and 1980s** ~ Another six records from fens and washes. *Bircham.*

CHESHIRE 1955, Dec 25 ~ A bird over Caldey Hill. *Cheshire and Wirral Ornithological Society* • **1968, Jan 15** ~ A female which had escaped from captivity at High Lane on 5th January was shot at Disley. *Cheshire and Wirral Ornithological Society* • **1968, Oct 16** ~ Female at Rostherne "appeared to have something on its legs, probably bells as used in falconry". *Cheshire and Wirral Ornithological Society*

CORNWALL 1838, Aug ~ With the exception of Polwhele, who probably referred to the Peregrine,* Bellamy is the only writer we are acquainted with who has recorded the occurrence of the Goshawk, *Astur palumbarius*, in Cornwall. He states (p. 198) that a young bird of this species was shot near Falmouth in August 1838. It is to be observed, however, that Mr. W. P. Cocks of Falmouth, in his "Contributions to the Fauna of Falmouth" ("Naturalist," 1851), does not include this bird in his list. Polwhele mentions the two species by name, referring to it as "scarce" in Cornwall; but cites no evidence of its occurrence. It has not unfrequently been confounded with the Peregrine, e.g., by Low, "Fauna Orcadensis". *Rodd.* **1951, Sept 10** ~ One flew over St Agnes. *Penhallurick.* • **1960, Sept 20** ~ A tired immature, reluctant to fly, was watched at close quarters by R. Khan in the Roseland. *Penhallurick.*

CUMBRIA 1800s ~ The Goshawk is an accidental visitant from continental Europe. An immature specimen was shot some years since near Penrith (T. Hope, MS.), and is preserved at Edenhall. *MacPherson and Duckworth* • **1983** ~ First recorded positive breeding success... a pair raised two young. *Hutcheson*

DERBYSHIRE Late Pleistocene ~ Sub-fossil remains from Robin Hood's Cave, Creswell (about 12,000 to 10,000 years ago). *Frost* • **Late Neolithic** ~ Recovered from Dowel Cave, Earl Sterndale (about 4,000 years before present). *Frost* • **1800s** ~ One was shot at Ashover, presumably last century (Whitlock). *Frost* • **1877** ~ There is a Goshawk in the Rolleston Hall collection, which was shot on the estate by one of the keepers in 1877, only a very short distance beyond the county boundary. *Whitlock* • The portions of the Testa de Neville relating to Derbyshire contain several entries referring to lands held by service of Goshawks. *Whitlock* • **1893** ~ One was seen at Bakewell (VCH). *Frost* • **1950s** ~ A female with a brood patch was found on a gamekeeper's gibbet

---

* In all probability the Peregrine, not mentioned by this name, is the bird referred to by Polwhele. Bellamy clearly distinguishes the two species [*Natural History of South Devon*].

on an estate in southern Derbyshire (T. W. Tivey, pers comm.). *Frost* • **1960s** ~ Breeding has been confirmed since at least 1966, and may have occurred for several years before that.

DEVON 1830 ~ Dr. Edward Moore has related that a Gos-Hawk's nest existed at South Tawton on the borders of the Forest in 1830, and that one of the birds was shot upon the nest. This bird was taken to Bolitho, the bird-preserver of Plymouth, and was sold by him to a Col. Burton. When interrogated by Mr. J. Gatcombe, Bolitho insisted that he had had a Gos-Hawk to stuff many years ago, and that he saw another hanging up in a gamekeeper's "larder" a few years since, which was too far decayed to do anything with. Now we are confident that no dependence is to be placed on these stories. *D'Urban*

DORSET 1888 ~ Pulteney speaks of it as not uncommon in his time, when game-preserving was not attended with such continuous persecution of birds of prey as at the present day, and when a larger extent of uncultivated country favoured its existence. *Mansell-Pleydell* • Pulteney writing at the end of the 18th century states "In Dorset not very uncommon, lives in the woods and frequents the furze and brakes," but possibly he confused it with other species such as harriers. *Blathwayt* • **1913, Nov** ~ Described as being "not very uncommon" at the end of the 18th century, it subsequently suffered considerably from persecution by gamekeepers and loss of habitat. The Goshawk's status in the county during the 19th century and the first half of the 20th century is vague with one shot at Canford about 2nd November 1913 the only definite report. *Green* **1100s and 1200s** ~

References of some interest, which suggest that this bird may not have been so rare in the 12th and 13th centuries, will be found in the fines of that period. *Glegg*

DURHAM 1872 ~ One well-authenticated record, a mature female which had been shot near Castle Eden Dene. *Barrie and Newsome.* • **1884**, January. A specimen of an immature female in the collection of Birmingham Museum. *Barrie and Newsome.* • **1934, Jan 2** ~ The immature bird that flew off the sea and up the Tees Estuary (cites Almond et al, 1939). *Barrie and Newsome.*

ESSEX 1100s and 1200s ~ References in the Court 'fines' of the 12th and 13th centuries there were a surprising number of references suggesting that Goshawks were probably relatively common (Glegg) *Wood* • **1619** ~Edmund Bert of Collier Row successfully trained Goshawks and wrote a book entitled An approved treatise of Hawks and Hawking. *Wool.* • **1822** ~ A female with jesses was shot at Audley End. *Wood* • **1891 and 1898** ~ Glegg, without explanation, dismissed records at Potton Island in 1881 and Hutton Hall around 1898. *Wood* • **1895** ~ The Rev. M. A. Mathew records that the last one known to him was seen by his brother, M. G. F. Mathew, R.N., near Harwich, one day in the winter of 1895. This is the only record which bears the stamp of reliability. *Glegg*

GLOUCESTERSHIRE 1903 ~ A pair of Goshawks built a nest close to Cheltenham, a singular and curiously interesting event ... the eggs were taken and identified at the Natural History Museum. *Hudson* • **1904** ~

The birds returned and nested at the same spot. This time both were shot by the owner of the land and the remains sent to a local bird-stuffer for what is ironically termed preservation. "It was a dastardly act," wrote Mr. Charles Witchell, who relates the facts. *Hudson.* **1831** ~ one *Mellersh.* **1878** ~ A pair Cirencester Woods. *Mellersh.*

HERTFORDSHIRE   1879, Sept 6 ~ Sage (1859) listed only one record, in square brackets, of one said to have been shot at Northaw. *Gladwin and Sage* • **1967, July 11** ~ One was seen in the act of taking a Wood Pigeon at Mardock Mill, Wareside, and what was probably the same bird wa1s seen again on 12 and 19 August. Gladwin and Sage • **1970, 25 Jan** ~ One was seen at Broxbourne Woods. *Gladwin and Sage* • **1980, Sept 24** ~ A female was seen near St Albans. *Gladwin and Sage*

JERSEY   1948, September ~ Pile and Long 18.9.1958. *Pile and Long.*

KENT   1950 ~ Harrison cites five definite records and only one this century on 7th October 1950. *Kent Ornithological Society*

LANCASHIRE   1863 ~ One was shot near Colne according to Mr. Henry Whalley, and Dr. Skaife writes as follows in the Mag. Of Nat. Hist., 1838, "Very rare, though shot or caught occasionally in the Forest of Bowland. A relative of my own has a beautiful pair, male and female, caught in a trap there a few years since." *Mitchell*

LEICESTERSHIRE AND RUTLAND 1881 ~ Probably extinct in the counties. Harley wrote:- "As regards the distribution of the Gos-Hawk in Leicestershire, I may remark that it used to occur not unfrequently in our woodlands and forest wilds, but of late years it has become exceeding rare. I have known it to be captured at Oakley, and the woods at Gopal, by both trap and gun," and further added that he had seen one, shot in Oakley Wood by a gamekeeper named Monk. In the 'Midland Naturalist,' 1882, p.62, Mr. Macaulay writes:- "One was seen in Allexton Wood in 1881"; but his informant, Mr. Davenport, replying to my enquiries, stated that this was a misconception of a verbal communication, and that, so far as he could recollect, "the taxidermist at Billesdon (Potter by name) had in his shop, for six or seven years (if not more), a bird shot at Allexton by a Mr. Brewster, who once lived at Allexton Hall: this bird was said to be a Gos-Hawk." Potter, on being written to, confirmed this, but having since then seen him, he informed me that the gentleman was in America. I am still in doubt whether a large female Sparrowhawk has not done duty in this, as in many similar cases, for the Gos-Hawk. *Browne* • **1856** ~ A male was shot near Barrowden (Rutland) in 1856, and is now in the possession of Dr. William Bell, of New Brighton, who believes he also saw a second in the same locality in the same year. *Haines*

LINCOLNSHIRE   1830, Dec ~ Bird shot or trapped near Louth. *Smith and Cornwallis* • **1864** ~ Pair attempted to nest in Normandy Park near Scunthorpe in 1864 but the female was shot. • **1871, May 23** ~ Bird shot or trapped near Louth. *Smith and Cornwallis* • **1910, April 29** ~ Bird shot or trapped near Louth. *Smith and Cornwallis* • **1919, Oct 5** ~ One near Normandy Park. *Smith and Cornwallis* • **1935, Sept 14** ~ One at Miningsby, near Spilsby. *Smith and*

Cornwallis • **1958, Feb 16** ~A vagrant recorded once [Gibraltar Point]. *Hickling and Davis*

NORFOLK   1832, summer ~ Immature Goshawk, shot at Yarmouth, sold by Smith in London for 7 pounds. *Lubbock* • **1833** ~ Fine specimen in eastern Norfolk. *Lubbock (cites Pagets)* • **1838** ~ Immature Goshawk killed at Marlingford by father of Mr Drake, who informed a Mr Stevenson. *Lubbock* • **1841** ~ Adult male killed at Colton. *Stevenson* • **1866** ~ Only mature plumage Gos known with certainty to have been killed in Norfolk, according to Lubbock. *Lubbock* • **1842** ~ Young female killed at Hingham ... probably the only example in mature plumage known with certainty to have been killed in Norfolk. *Stevenson* • **1843** ~ About November, Goshawk brought in to Yarmouth after lighting in the rigging of a ship • **1850, Nov** ~ Young bird killed at Stratton Strawless. *Stevenson* • **1851** ~ Immature female shot near Norwich whilst preying on a hare. *Stevenson* • **1854, April** ~ Immature male in very beautiful plumage taken at Catfield. *Stevenson* • **1858, Nov 23** ~ A female in its first year's plumage killed at Hempstead. *Stevenson* • **1863, autm** ~ Young bird trapped at Riddlesworth. *Stevenson* • **1869, Dec 27** ~ Immature killed at Beeston Regis, now in collection of Mr Cremer. *Lubbock* • **1869, Dec 5** ~ Immature shot at Filby, now in collection of Rev C. J. Lucas. *Lubbock* • **1876, early** ~ Immature killed near Melton Constable. *Lubbock* • **1881** ~ Influx of Sparrowhawks noted, mostly females, some dead on tide-line. *Paterson* • **1886** ~ Female landed on a fishing boat. *Paterson* • **1901, March 9** ~ Female caught in a rabbit-trap at Weybourne and is now in the

Connop collection. *Pashley* • **1942** ~ Goshawk at Hickling in November. *Bannerman* • **1942** ~ Goshawk at Dereham in December. *Bannerman* • **1952** ~ Goshawk seen flying in from North Sea by R. A. Richardson at Cley Bird Observatory in June. *Bannerman*

NORTHUMBRIA   1752 ~ Dr. James Hill in his History of Animals (1752) speaks of the Goshawk nesting in Northumberland. *Hudson* • **1833, Oct** ~ Immature Goshawk shot at Bellingham • **1841, Feb 18** ~ "Fine female" shot at Bolam Bog • **1841** ~ Female at Alnwick • **1844** ~ Adult female taken in a trap near Bedlington. *Bolam* • **1845, April** ~ Young female killed at Woodburn, in Redewater. *Bolam* • **1849, Oct** ~ Young male killed at Kielder on the North Tyne. *Bolam* • **1854, Feb** ~ Young male killed at Whickhope, on the North Tyne. *Bolam* • **1862** ~ Adult taken in Alnwick Park. *Bolam* • **1862** ~ Mr. Hughes, of Middleton Hall, has an immature specimen which was trapped near the top of Cold Law. *Bolam* • **1876, Jan 16** ~ Young female shot by T. Elliott, gamekeeper at Lilburn Tower. *Bolam* • **1876, Jan** ~ Another, which escaped, was seen to clutch a partridge a short time afterwards, and remained in the woods for several weeks; this might possibly be the same bird which was seen by one of the Chillingham gamekeepers, in Trickley Wood, in the following March. *Bolam, 1912* • **1876, Jan** ~ One was shot at Benwell, near Newcastle. *Bolam* • **1897, early** ~ A fine immature female was trapped at Middleton Hall, Belford, and is preserved at Ayton Castle. *Bolam* • **1897** ~ Mr Henry Liddell-Grainger, who was at the time occupying Middleton Hall, added that when he was shooting in Detchant Wood, one of his men, who was acting as a stop, told him of "a most awful-like bird, many times bigger than a hawk," which had passed him, and which was doubtless another Goshawk. *Bolam* •

SCILLY ISLES 1935, December 28 ~ Adult US race killed at Tresco by F. W. Frohawk. *Penhallurick.*

SUFFOLK 1833, March 16 ~ Hoy, who recorded that an adult male was trapped on a Red-legged Partridge by Rowley's keeper at Stoke-by-Nayland. Hoy thought that it might have been an "escape". *Ticehurst* • In a trap baited with a red-legged partridge. *Babington* • **1841, Jan** ~ Fine adult male shot by Mr. Spalding in a wood at Benacre; now in possession of Mr E. Spalding of Middleton. *Babington* • **12 Jan** ~ Described as an immature, killed in a wood. *Lubbock* • **1843** ~ One alighted in the rigging of a ship brought to that place [Yarmouth]. *Babington* • **1849, Nov 20** ~ Fine male in immature plumage shot at Westhorpe and brought in the flesh to Dr. Bree, who has it. Seen at Bacton 1881. *Babington* • **1854, Nov or Dec** ~ One killed at Elveden. *Babington* • **1859, Jan** ~ An immature female shot at Somerleyton by Mr John Gould; now preserved at the Hall. *Babington* • **1866** ~ One at Barrow found with a small trap on its foot. *Babington* • **1868, Dec** ~ One shot at Rendlesham. One shot at Butley about the same time. *Babington* • **1868** ~ One procured at Trimley. *Babington* • **1874** ~ Record from Saxham. *Payn* • One killed at Aldeburgh; in possession of Col. Thelluson; and another immature female killed at the same place (Bury Museum). *Babington* • Two young birds killed on the Rendlesham estate; in Lord Rendlesham's Collection. *Babington* • **1886** ~ Taken on boat in the North Sea. *Babington* • **1893, March 2** ~ Adult

female taken at Somerleyton on board a fishing boat. *Ticehurst; Paterson* • Recorded also well inland as at Elveden, Westhorpe near Stowmarket, Saxham and Barrow. *Ticehurst; Paterson* • It might be thought that the great decrease of this bird in East Anglia might be connected with a corresponding decrease in Scandinavian breeding birds but this apparently is not so, as the late Professor Collett said that in 1871–1890 head-tax was paid on an average of 3,764 Goshawks yearly in Norway alone. In 1913–1917 the annual slaughter had risen to 4,666 birds, though probably Buzzards are included in this total; in 1920-1926 the annual toll was 3415. One can only suppose, therefore, that its migrations very occasionally reach our shores. *Ticehurst; Patterson* • **1946, Oct** ~ An immature male was shot at Saxham. *Payn, citing Sir C. Magnay, Bt* • **1962** ~ Since 1950 some eighteen to twenty goshawks have been reliably reported, probably reflecting the greater number of observers on the lookout for them. Most occurrences were on or near the coast and consisted of single birds but two were seen at Benacre in September 1957 and one shot two days later in north Suffolk was probably one of them. *Payn* • **1958, Feb** ~ A pair were seen displaying in Breckland • **1959** ~ One was seen drinking at a water-tank [Breckland]

SUSSEX 1920s/1930s ~ Breeding • **1951** ~ Final breeding. *Harrison*

WEST MIDLANDS 1853, 1857 and 1877 ~ Smith listed three records of the Goshawk, at Swythamley in 1853, Uttoxeter in 1857 and Rolleston in 1877. *West Midlands Bird Club*

WILTSHIRE Introducing the goshawk as a bird of Wilts on the authority of the Rev. A. P. Morres, who gives good and substantial evidence from the mouth of Captain Dugmore, a gentleman who seems specially qualified to pronounce an opinion, that while hawking in the meadows near Salisbury, the tame goshawk on his wrist showed by its manner and cry that a wild bird of the same species was at hand; and having his attention thus aroused, he clearly saw a wild Goshawk flying in a straight course high over his head, and he added that he had no doubt as to the bird's identity, since he was so very familiar with it from constantly hawking with the same species. Moreover I see no reason why the goshawk should not occasionally visit us, seeing how common it is in Germany, where I have fallen in with it more than once; and how capable it is of prolongéd flight. *Smith*

WORCESTERSHIRE   1954 and 1958 ~ Birds observed in Worcestershire in 1954 and 1958 were considered to be of this species by experienced observers

YORKSHIRE   1825 ~ The only Yorkshire specimen on record was shot at Cusworth by Mr. Wrightson's gamekeeper. *Allis* • 1852, Oct 15 ~ A male and female were killed by Mr. G. S. Gibbs, gamekeeper to Mr. H. Kirk of Stockton-on-Tees. *Nelson* • 1852, Feb ~ Goshawk at Driffield • 1850s ~ Adult female taken by Lady Downe's keeper at Wykeham. *Nelson* • 1863 ~ This spring the nest of this rare bird was found in some ivy which surrounds an old oak tree which is situated in the boundary hedge of a plantation. I did not see the nest, but the eggs were of a very pale blue. They were unfortunately broken by one of the possessor's children. *Ranson, J. in The Zoologist, 1863* • 1864, March 2 ~ At the meeting of the Yorkshire

Naturalists' Club, D. Graham exhibited a fine specimen taken near Oswaldkirk by Mr. J. Bower. *Nelson* • 1864, June ~ Young male in singular 'cuckoo' plumage, trapped on the Lockton Moors, near Pickering (mistaken for an Iceland falcon). *Nelson* • 1864, first week Oct ~ Mr. R. Lorrimer obtained a fine specimen whilst in pursuit of its prey on Filey Brigg. *Nelson* • 1871 ~ Adult female trapped on a rabbit warren near Harrogate, April 15. *Nelson* • 1875 ~ One was reported at Ewecote, near Whitby, Aug 29. *Nelson* • 1877 ~ A Goshawk was captured at Ewecote, by Mr. T. Crosby, for whom it was stuffed by Mr. Kitching of that town. Possibly these last two records related to the same example. *Nelson* • 1877, Jan 23 ~ Mr. Matthew Bailey of Flamborough has supplied the particulars of a specimen obtained near Flamborough; from this communication the following account is condensed: The bird, a fine old female, had frequented the neighbourhood for some weeks, baffling all attempts made to shoot it, until it was observed by the gamekeeper of the Rev. Y. G. Lloyd-Greame, of Sewerby Hall, to kill a full-grown rabbit, which it had carried about twenty yards when he shot at but missed it. Concealing himself in an adjoining wood the keeper had not long to wait, as the bird soon returned and was killed. This bird, Mr. Bailey informs me, is now in the collection of Sir Vauncey Crewe, Bart. *Nelson* • 1889, Feb 14 ~ Adult male trapped by one of the keepers at Keldy Castle, near Levisham. *Nelson* • 1893, early May ~ An adult female Goshawk was shot at the nest by Mr. W. M. Frank, keeper to Capt. Duncombe at Westerdale, in Cleveland. Mr. Frank was under the impression she had a mate, but he did not see two birds together. Two eggs were sent to the Norwich Museum, along with the parent bird. A suggestion was put forward by Mr. Heatley Noble, who supplied the fact to Mr. Southwell, that the Hawk may have been an escaped trained bird. *Nelson.* • **1896, Sept 27** ~ A pair were reported at Easington, near Spurn, by the late H. B. Hewetson. *Nelson* • **1896-97, winter** ~ A specimen captured at Escrick was preserved by Mr. J. Pulleine of Selby. *Nelson* • **1897, Dec** ~ An immature male occurred at Wheeldale, near Whitby, and is now in the possession of Mr. J. C. Walker of that town. *Nelson* • **1897** ~ Nelson published records of about 20 Goshawks in Yorkshire between 1825 and 1897, most of which were inevitably shot or trapped. *Mather* • **1899** ~ Some of the attempts to procure birds of prey were persistent and ruthless, as shown in the account of one at Flamborough which Matthew Bailey communicated to Nelson. *Mather* • **1906** ~ A juvenile shot at Filey in January 1906 is in the Dorman Museum, Middlesbrough. *Mather* • **1939, Jan 9** ~ A bird was seen near to Runswick Bay by W. S. Medlicott, who stated: "I watched this bird hunting for some minutes. Having kept several for hawking purposes and trained them, I have no doubt whatever as to its identity." *Chislett* • **1949, Nov 1** ~ P. Young, a gamekeeper-naturalist, who knows the raptorials well, had good views of a bird at Ilton ... a Carrion Crow was following the Goshawk. *Chislett*

IRELAND   1600s, first half ~ Goshawks from the north of Ireland were the best in the world, according to Sir James Ware. *Hudson* • **1686** ~ Tyrone Goshawks are cited in Richard Blome's The Gentleman's Recreation (1686), as, with those from Muscovy and Norway, the best for sport. *Hudson* • **1844** ~ Shot by Lord Meath's gamekeeper, recorded as seen fresh by Dr J. R. Kinahan, at Kilruddery, Wicklow. *Kennedy* • **1846, autumn** ~ Immature male obtained at Longford. *Kennedy. Also in Watters' Birds of Ireland, 1853* • **1870** ~ Immature male seen in March, April or May at Ballymanus Wood, Wicklow. *Zoologist, 1870.* • **1870** ~ Adult female American race Goshawk taken at Parsonstown, Kings County, in spring. *Bent, Smithsonian* • **1870** ~ Female American race Goshawk shot and presented to the museum. Galtee Mountains, Tipperary. *Kennedy* • **1870** ~ American race Goshawk shot at Offaly. *Kennedy* • **1935** ~ Adult female trapped at Carnakelly Bog, Athenry, Galway on December 23. *Kennedy* • **1955** ~ Goshawk reported seen on April 29, Roscarbery, Cork. *Kennedy* • **1959** ~ Adult Goshawk reported seen on August 7 at Blennerville, Kerry. *Kennedy* • **2000** ~ There are only some 25 Irish records of *gentilis* (Palmer 2000). *Robinson*

NORTHERN IRELAND   1919 ~ Adult male American race Goshawk obtained on February 24 at Strabane, Tyrone. Now in the museum. *Kennedy* • **1956** ~ Goshawk reported seen on August 12 at Lough Beg. *Kennedy* • **1956** ~ Goshawk reported seen on September 16 at Lough Beg. *Kennedy* • **1956** ~ Goshawk reported seen at Duncrue Street Marsh, Belfast on Oct 1. *Kennedy*

SCOTLAND   1684 ~ Sibbald includes it in his List. *Baxter and Rintoul* • **1769** ~ Pennant said that it bred in the forests of Invercauld. *Baxter and Rintoul* • **1795** ~ Campsie, in

Stirlingshire: "The Goshawk, which builds its nest upon trees in sequestered places, is likewise native of this parish." This record was made by the Rev. John Lapslie, a naturalist of some repute. *Baxter and Rintoul* • 1784 ~ Colonel Thornton, in his Sporting Tour, says that in "the forest formed by Glenmore and Rothiemurchus... are also some eyries of Goss-hawks, some of which we saw." *Baxter and Rintoul* • 1825 ~ One obtained in Forfarshire and seen by Fenton, a well-known taxidermist of Edinburgh. *Harvie-Brown* • 1835 ~ Record from Macgillivray, quoted by R. Gray. *Harvie-Brown* • 1842 ~ I have the record of a Goshawk killed in Kemback Wood, near St. Andrews. *Harvie-Brown* • 1849 ~ St. John wrote "The only pace where I know of it breeding regularly is in the forests of Darnaway; but I am told that they also breed in the large firwoods of the Spey." *Baxter and Rintoul* • 1850s ~ Still breeding in Inverness-shire. *Harrison* • 1859, April ~ Reported seen at rabbit warren by Saxby at Balta, Shetland. *Saxby* • 1860 ~ Female shot in winter at Skaw, Unst. Acquired by Saxby soon after. *Saxby* • 1862 ~ Said by St. John to still be breeding at Glenmore Forest. *Bannerman* • 1865 ~ More says that the Goshawk used to breed in the woods of Castle Grant and in the woods of Dulnan, when W. Dunbar said he had taken the young out of the nest. *Baxter and Rintoul* • 1866 ~ R. Gray tells of one "at Glamis about five years ago". It had been caught in a pole-trap. *Harvie-Brown* • 1869 ~ American race Goshawk shot at Schiehallion, Perthshire in May. Kept by Robert Gray and seen several times by Harvie-Brown. *Baxter and Rintoul* • 1871 ~ Two eggs in the Hargitt collection are said to have been taken

at Balmacara in April. *Baxter and Rintoul* • 1871 ~ Robert Gray in his Birds of West Scotland (1871) writes: "Within a comparatively recent period I have known the Goshawk breed in Kirkcudbrightshire"; and he adds that a pair of Ravens were driven from their nest by Goshawks, who appropriated it to their own use. *Hudson* • 1877 ~ One certainly was obtained at Elie (Forth) and I have seen a sketch of it done by Mr. W. Evans from the bird when it was in the possession of Mr. Small, Edinburgh. *Harvie-Brown* • 1879, June ~ Last native pair shot at Rohallion, near Birnam, Perthshire, by gamekeeper J. Boath. *Perth Museum* • 1883 ~ "The last native pair probably bred at Rohallion, near Birnam" (Perth and Kinross) [this is when the stuffed pair shot in 1879 were obtained by Millais]. *Scottish Ornithologists' Club, 2007*

WALES 1800 ~ Last recorded nest found at the head of the Neath Valley about 1800. *Lovegrove, Williams and Williams* • Early 1800s ~ A very unsatisfactory record, now a century and a half old, square bracketed in previous lists. *Cardiff Naturalists' Society, 1967* • 1800s ~ Bird shot on Mostyn Estate mentioned by Forrest (1907). *Lovegrove, Williams and Williams*, • 1962, Aug 30 and Sept 20 ~ Sightings, almost certainly of the same bird, over the Kenfig dunes. *Cardiff Naturalists' Society, 1967* • 1969, Aug 10 ~ First confirmed breeding when an adult with a fully-fledged youngster was seen in the Pembrey Forest. *Lovegrove, Williams and Williams*

UK 1847 ~ St. John speaks of the Goshawk as "now nearly extinct in this country, whereas a few years ago it bred regularly". *Hudson*

Bibliography and further reading

Avery, M. 2012. *Fighting For Birds*. Pelagic Publishing, Exeter.

Avery, M. and Leslie, 1990. *Birds And Forestry*. Poyser, London.

Babington, C. 1884–1886. *Catalogue of the Birds of Suffolk*. John Van Voorst, London.

Baker, J. A. 1967. *The Peregrine*. New York Review Books, New York.

Bannerman, D. A. and Lodge, G. E. 1953–1963. *Birds of The British Isles*. Oliver and Boyd, Edinburgh.

Baxter, E. V., and Rintoul, L. J 1953. *The Birds of Scotland*. Oliver and Boyd, Edinburgh.

Bell, I. 1993. *Robert Louis Stevenson: Dreams of Exile*. Headline, London.

Bent, A. C. 1937-1938. *Life Histories of North American Birds of Prey* (Volume 1-2). Dover. New York.

Bent, A. C. *Life History Series* Smithsonian

Bewick, T. 1797. *A History of British Birds*.

Bijleveld, M. 2004. *Birds of Prey in Europe*. Macmillan, London.

Bircham, P. M. M. 1989. *The Birds of Cambridgeshire*. CUP.

Blathwayt, F. L. 1933. *A Revised List of the Birds of Dorset*. Dorset Natural History and Archaeological Society.

Bolam, G. 1912. *Birds of Northumberland and the Eastern Borders*. Henry Hunter Blair, Alnwick.

Booth et al 1984. *The Birds of Orkney*. Orkney Press, Orkney.

Bowie, K. and Newsome, M. Eds. 2012. *The Birds of Durham*. Durham Bird Club.

Brigham, R. 1988. *Guns and Goshawks*. Blandford, London.

Brooks, P. 1972. *The House of Life – Rachel Carson at Work*. Houghton Mifflin. Boston.

Brown, L. 1976. *British Birds of Prey*. Collins New Naturalist Series, London.

Brown, P. 1964. *Birds of Prey – A Survival Guide*.

Browne, B. 1889. *The Vertebrate Animals of Leicestershire and Rutland*. Midland Educational Company Limited, Birmingham and Leicester.

Cairns, P. and Hamblin, M. 2007. *Tooth and Claw – Living Alongside Britain's Predators*. Whittles Publishing, Dunbeath.

Carson, R. 1962. *Silent Spring*. Houghton Mifflin, Cambridge Massachusetts.

Chislett, R. 1950. *Yorkshire Birds*. A. Brown & Sons, Limited, London.

Christy, M. 1890. *The Birds of Essex*. Edmund Durrant & Co, Chelmsford.

Cocker, M. and Fanshawe, J. Eds. 2011. *The Peregrine*, the Hill of Summer *and Diaries: The Complete Works of J. A. Baker*. Collins, London.

Collings, M. 2007. *A Very British Coop. Pigeon Racing from Blackpool to Sun City*. Macmillan, London.

Cramb, A. 1996. *Who Owns Scotland Now? The Use and Abuse of Private Land*. Mainstream Publishing, Edinburgh and London.

Crofts, R. and Boyd, I., Eds. 2005. *Conserving Nature: Scotland and the Wider World*. John Donald, Edinburgh.

D'arcy, G. 1999. *Ireland's Lost Birds*. Four Courts Press, Dublin.

D'Urban, W. S. 1892. *The Birds of Devon*. Porter, London.

Dazley, R. A. and Trodd, P. 1992. *The Birds of Bedfordshire*. Bedfordshire Natural History Society.

Deans, P. et al, Eds. 1992. *An Atlas of the Breeding Birds of Shropshire*. Shropshire Ornithological Society.

Dick, D. 2012. *Wildlife Crime*. Whittles, Dunbeath.

Dobbs, A. 1975. *The Birds of Nottinghamshire Past and Present*. David & Charles, Newton Abbot.

Dutt, W. A. 1906. *Wild Life in East Anglia*. Methuen, London.

Fox, N. 1995. *Understanding the Bird of Prey*. Hancock House, Canada.

Frost, R. A. 1978. *Birds of Derbyshire*. Moorland Publishing Company.

George, J. 2004. *My Side of the Mountain*. Puffin, London.

Gibbons, D. W. Reid, J. B., Chapman, R. A. 1993. *The New Atlas of Breeding Birds in Britain and Ireland: 1988–91*. T. & A. D. Poyser, London.

Glasier, P. 1978. *As the Falcon her Bells*. Futura Publications Limited, London.

Gladwin, T. W. and Sage, B. L. 1986. *The Birds of Hertfordshire*. Castlemead Publication, Ware.

Glegg, W. E. 1929. *A History of the Birds of Essex*. Witherby, London.

Green, G. 2004. *The Birds of Dorset*. Christopher Helm. London. Guest, J. P. et al 1992. *The Breeding Bird Atlas of Cheshire and Wirral*. Cheshire and Wirral Ornithological Society.

Haines, C. R. 1907. *Notes on the Birds of Rutland*. R. H. Porter, London.

Hamilton Lytle, M. 2007. *The Gentle Subversive: Rachel Carson, Silent Spring and the Rise of the Environmental Movement*. Oxford University Press, USA.

Hancock, J. 1874. *The Birds of Northumberland and Durham*. Williams and Norgate, London.

Harrison, C. 1988. *The History of the Birds of Britain*. Collins, London.

Harrison, G. R. 1982. *The Birds of the West Midlands*. West Midland Bird Club.

Harvie-Brown, J. A. 1906. *A Vertebrate Fauna of Scotland. The Tay Basin and Strathmore*. David Douglas. Edinburgh.

Heathcote, A. et al. Eds. 1967. *The Birds of Glamorgan*. Cardiff Naturalists' Society.

Hedley Bell, T. 1962. *The Birds of Cheshire*. John Sherratt and Son, Altrincham.

Hinde, T. 1985. *Forests of Britain*. Victor Gollancz Ltd, London.

Hosking, E. 1970. *An Eye for a Bird*. Hutchinson.

Hudson, W. H. 1915. *Birds and Man*. Duckworth & Co, London.

Hudson, W. H. 1923. *Rare Vanishing and Lost British Birds* compiled from notes by Linda Gardiner. J. M. Dent & Sons Ltd. London and Toronto.

Hudson, W. H. 1987. *The Illustrated Shepherd's Life*. Guild Publishing, London.

Hutcheson, M. 1986. *Cumbrian Birds*. Frank Peters (Publishing) Ltd, Kendal.

Jensen, D. 2008. *A Language Older than Words*. Finch Publishing, Sydney.

Jones, G. L. and Mabey, R. 1993. *The Wildwood*. Aurum Press, London.

Joynt, G. et al, Eds 2008. *The Breeding Birds of Cleveland*. Teesmouth Bird Club.

Kennedy, P. G. 1961. *A List of the Birds of Ireland*. Stationery Office, Dublin.

Kenward, R. 2005. *The Goshawk*. T. &A. D. Poyser, London.

Kenward, R. E. and Lindsay, I. M. 1981. *Understanding the Goshawk*. International Association of Falconry and Birds of Prey.

Lever, C. 2005. *The Naturalised Birds of the World*. T. & A. D. Poyser, London.

Lovegrove, R. 2007. *Silent Fields: The Long Decline of a Nation's Wildlife*. Oxford University Press, Oxford.

Lovegrove, R., Williams, G. and Williams, I. 1994. *Birds in Wales*. T. & A. D. Poyser, London.

Lubbock, 1879. *Observations on the Fauna of Norfolk*. Jarrold.

MacPherson and Duckworth. 1886. *The Birds of Cumberland*. Thurnam & Sons, Carlisle.

Mansell-Pleydell, J. C. *The Birds of Dorsetshire*. R. H. Porter, London.

Mather, J. R. 1986. *The Birds of Yorkshire*. Croom Helm, London.

Mawson, G. 'Apparent Nesting Association of Northern Goshawks and Firecrests'. British Birds, 103 April 2010 243–247.

Mccurran et al. 2008. *The 2nd Atlas of Breeding Birds in New York State.*

Mcvean, J. D. and Lockie, D. N. 1969. *Ecology and Land Use in Upland Scotland.* Edinburgh University Press, Edinburgh.

Mearns, B. and R. 2002. *The Bird Collectors.* T. & A. D. Poyser, London.

Mellersh, W. L. 1902. *A Treatise on the Birds of Gloucestershire.* R. H. Porter, London.

Mitchell, F. S. 1892. *The Birds of Lancashire.* Gurney and Jackson, London.

Morrison, M. L. 2006. *The Northern Goshawk: A Technical Assessment of its Status, Ecology, and Management.* Cooper Ornithological Society.

Nelson, T. H.. 1907. *The Birds of Yorkshire.* A. Brown & Sons Ltd, London.

Newton, I. 2010. *Population Ecology of Raptors.* T. & A. D. Poyser, London.

Newton, I. 2010. *The Sparrowhawk.* T. & A. D. Poyser, London.

Nicholson, E. M. 1926. *Birds in England.* Chapman and Hall, London.

Nurse, A. 2013. *Animal Harm.* Ashgate.

Oakes, C. 1953. *The Birds of Lancashire.* Oliver and Boyd, Edinburgh.

Orton, D. A. 1989. *The Hawkwatcher.* Collins, London.

Palmer, E. M. 1968. *The Birds of Somerset.* Longman.

Parkin, D. T., Knox, A. G. 2010. *The Status of Birds in Britain and Ireland.* Christopher Helm, London.

Parry-Jones, J. 2003. *Falconry.* David & Charles, Devon.

Parslow et al. 1971. *The Status of Birds in Britain and Ireland.* Blackwell Scientific Publications, Oxford.

Pashley, H. N. 1925. *Notes on the Birds of Cley, Norfolk.* H. F. & G. Witherby, London.

Paton, E. R. and Pike, O. G. 1929, 2002. *The Birds of Ayrshire.* Castlepoint Press.

Payn, W. H. 1962. *The Birds of Suffolk.* Ancient House Publishing, Ipswich.

Penhallurick, R. D. 1978. *The Birds of Cornwall and the Isles of Scilly.* Headland, Cornwall.

Petersen, R., Mountfort, G., Hollom, P. A. D. 1965. *A Field Guide to the Birds of Britain and Europe.* Collins, Great Britain.

Pile, C. G. and Long, M. L. 1959. *A Check List of the Birds of Jersey.* Societé Jersiase. Jersey.

Rackham, O. 1993. *The History of the Countryside.* J. M. Dent, London.

Rodd, E. H. 1880. *The Birds of Cornwall and the Scilly Islands*. Trubner and Co. London.

RSPB, 2012. *Birdcrime* 2011. RSPB, UK.

Rutter, E. M, et al. 1964. *A Handlist of the Birds of Shropshire*. Shropshire Ornithological Society.

Sage, B. L. 1959. *A History of the Birds of Hertfordshire*. Barrie and Rockliff, London.

Salvin, F. H. and Brodrick, W. 1980. *Falconry in the British Isles*. Windward, Leicester.

Samstag, T. 1988. *For the Love of Birds – The Story of the RSPB*. RSPB, Bedfordshire.

Saxby, H. L. 1874. *The Birds of Shetland*. Maclachlan & Stewart, Edinburgh.

Sharrock, J. T. R. 1976. *The Atlas of Breeding Birds in Britain and Ireland*. BTO/IWC, Great Britain.

Smith, A. C. 1887. *The Birds of Wiltshire*. R. H. Porter, London.

Smith, A. E. and Cornwallis, R. K. 1955. *The Birds of Lincolnshire*. Lincolnshire Naturalists' Union, Lincoln.

Smith, T. 1938. *The Birds of Staffordshire*. North Staffordshire Field Club.

Smout, T. C. 1997. *Scottish Woodland History*. Scottish Cultural Press, Edinburgh.

Smout, T. C. Ed. 2000. *Nature, Landscape and People since the Second World War*. Tuckwell Press, East Linton.

Smout, T. C. *Nature Contested: Environmental History in Scotland and Northern England Since 1600*. Edinburgh University Press, Edinburgh.

Snyder, N. 1992. *Birds of Prey*. Voyageur Press, USA.

Sterland, W. J. 1869. *The Birds of Sherwood Forest*. J. Reeve & Co

Sterland, W. J. 1879. *Descriptive List of the Birds of Nottinghamshire*. William Gouk, Mansfield.

Stevenson, H. 1866. *The Birds of Norfolk*. John Van Voorst, London.

Summers, G. 1975. *The Lure of the Falcon*. Fontana/Collins, Glasgow.

Taylor, D. W. et al, 1984. *The Birds of Kent*. Kent Ornithological Society.

Ticehurst, C. B. 1932. *A History of the Birds of Suffolk*. Gurney and Jackson, London.

Tomkins, S. 1989. *Forestry in Crisis: The Battle for the Hills*. Christopher Helm, London.

Toyne, E. P. 1994. *Studies on the Ecology of the Northern Goshawk* Accipiter Gentilis *in Britain* (Unknown binding). University Of London.

Trodd and Kramer. 1991. *The Birds of Bedfordshire*. Castlemead, Welwyn.

Ussher, R. J., Warren, R. 1900. *The Birds of Ireland*. Gurney and Jackson, London.

Venables, W. A. et al. *The Birds of Gwent*. Christopher Helm, London.

Wernham, C. et al, Eds. 2002. *The Migration Atlas*. T. & A.D. Poyser, London.

Whitaker, J. 1907. *Notes on the Birds of Nottinghamshire*. Walter Black & Co. Ltd, Nottingham.

White, T. H. 1938. *The Sword in the Stone*. Collins, London.

White, T. H. 1951. *The Goshawk*. New York Review of Books, New York.

White, T. H. 1965. *America at Last – The American Journal of T. H. White*. G. P. Putnam's Sons, New York.

Whitlock, F. B. 1893. *The Birds of Derbyshire*. Bemrose & Sons, Limited, London.

Whitlock, R. 1953. *Rare and Extinct Birds of Britain*. Phoenix House.

Winn, M. 1999. *Red-tails in Love*. Pantheon, New York.

Wood, S. 2007. *The Birds of Essex*. Christopher Helm, London.

Yalden, D. and Albarella, U. 2009. *The History of British Birds*. Oxford University Press.

1989. *Goshawks: Their Status, Requirements and Management* (Bulletin). Stationery Office Books, London.

Survival Special: *Goshawk – The Phantom of the Forest,* Anglia Television 1992. Producer Hugh Miles. Narrator Andrew Sachs. Shown 1992 on London Weekend Television, 1996 on Channel 4, 1999 on Channel 5

Video of Goshawk filmed at local gravel pits, 2010. **http://paxtonpits.blogspot.com/**

Further information on the history of Goshawks at **http://conorjameson.tumblr.com/go**

Index